Water Treatment Specification Manual

Water Treatment Specification Manual

Frank Rosa

Field Engineer, Metropolitan Refining Company,
Long Island City, New York

McGRAW-HILL BOOK COMPANY

New York   St. Louis   San Francisco
Auckland   Bogotá   Hamburg   Johannesburg
London   Madrid   Mexico   Montreal   New Delhi
Panama   Paris   São Paulo   Singapore
Sydney   Tokyo   Toronto

Library of Congress Cataloging in Publication Data

Rosa, Frank.
    Water treatment specification manual.

    Bibliography: p.
    Includes index.
    1. Feed-water—Purification.   2. Heating.
3. Air conditioning.   I. Title.
TJ379.R67 1985   697   85-5215
ISBN 0-07-053606-6

1234567890   DOC/DOC   898765

ISBN 0-07-053606-6

*The editors for this book were Joan Zseleczky and
Lester Strong, the designer was Elliot Epstein, and
the production supervisor was Teresa F. Leaden. It was
set in Century Schoolbook by Progressive Typographers.
Printed and bound by R. R. Donnelley & Sons Company.*

This book is dedicated to my beloved wife ELBA IRIS, *Action Speaks Louder Than Words,* and my wonderful children Efrain J., Julie E., and Christina E.

Contents

Preface

The objective of this book is to provide consulting firms, professional engineers, mechanical contractors and direct clients with the means for solving water treatment problems, while at the same time using minimal chemistry to illustrate its points.

Up till now, design engineers wanting sets of specs and drawings for projects requiring water treatment had to utilize sets already existing, altering them to meet the new conditions. The engineers may have also contacted water treatment firms to supply specifications for the proposed projects, which guidelines in all probability were very specific and tended to limit the engineers to the use of particular programs and at times unique chemical feed equipment unavailable elsewhere. But with the help of this book design engineers can now specify nonproprietary water treatment programs for clients and feel confident, if the specs are met, that the clients will be pleased. The drawings used to illustrate the manual, based on extensive field experience, try to eliminate those water treatment call-back problems that can sour engineer-client relationships.

For mechanical contractors, this work will help trace water treatment problems on troublesome jobs — that is, will help answer questions such as "Why are the heat pump coils plugging?" or "Why is a particular water treatment feed system not suitable for this project?"

Plant managers or physical plant directors for educational facilities can use this work to solve boiler flooding problems, upgrade chemical feed equipment, or publish a set of specs for comparative bidding on water treatment projects for the facility.

This manual can be effectively used in schools that teach HVAC courses since it is devoid of chemical terms which tend to confuse the HVAC and mechanical engineering student. Students in the HVAC field understand plumbing, pressures, thermodynamics, and associated HVAC terms. However, it is unlikely they will know the differences between phosphonic and phosphoric acids or will even care to know them. The HVAC student is interested in why a problem occurs, *not* in chemical terms but in terms that can be related to. For example, on finding a condenser loaded with lime, will HVAC technicians want to hear that the unit scaled up because the solubility limits of the carbonates of calcium and magnesium were exceeded, and this, compounded by high alkalinity, caused the carbonate salts to layer on the heat exchange surfaces of the condenser? Or will they want to know that the unit plugged because of insufficient bleed-off?

The reader will note that this work is devoid of references to specific products manufactured by water treatment firms. It was felt that their inclusion would take away rather than contribute to the value of the book. The engineers consulted during its writing felt there were enough puff works available without adding another.

The author is indebted to the following for assistance and inspiration:

Mr. Michael H. Johnson, plumbing designer for a major northeastern architectural firm, for his help in making the author's drawings suitable for publication.

The late Messrs. T. T. Peck, C. F. Schweizer, and W. J. Covney for the fine examples they set.

Clients, past and present, for allowing the author to run tests on their equipment.

Central New York ASHRAE members for their advice and encouragement.

Central New York consulting engineers for their repeated urgings to get this work published.

The entire staff at Metropolitan Refining Company, Inc., for the experience the author has gained there since 1969.

To Master Efrain J. Rosa for spotting numbering errors in the drawings.

Frank Rosa

Water Treatment Specification Manual

ONE

Water Treatment Justification

This chapter provides guidelines that will enable the reader to choose a heat rejector based on sound principles. When does one opt for an air-cooled condenser vs. a water-cooled unit? Which chemical feed system is best? Although many factors must be taken into consideration in choosing a particular heat rejector, the reader or design engineer should not overlook a prime parameter — the quality of the water supply! Let us analyze the water quality for each system under consideration and strive to use the information gleaned here in the decision-making process.

OPEN RECIRCULATING WATER SYSTEMS

The primary factor in choosing a water-cooled vs. an air-cooled condenser should be the quality of the water supply. However, design engineers are often confused by the term "quality" as it is applied to recirculating water systems. As used here, this term describes the ramifications of using the water for a particular purpose. For example, a water supply may be of excellent quality for drinking purposes but be of very poor quality for process use. In open recirculating water systems, one is concerned with those parameters which will impede heat transfer and thus affect heat-rejecting efficiency. The parameters of concern are *hardness,* which must not exceed 1200 parts per million (ppm) as calcium carbonate; *alkalinity,* which must not exceed 500 ppm as calcium carbonate; and *silica,* which must not exceed 150 ppm. These are the maximum parameter limits, and to exceed them invites problems. This is not to say that one will not experience problems as long as one does not exceed the maximums. Rather, as a primary analytical tool, one should take these parameters into consideration. Since the author does not intend to burden the reader with many chemical terms or formulas, the reader is urged to review other works on the subject.[1,8,9] However, the author does not feel that knowledge of chemistry is essential to the design engineer, just as a knowledge of mechanical engineering is not needed to operate an automobile. One need only understand that certain parameters are not to be exceeded to design successful water treatment facilities.

Let us now analyze some water supplies and examine their use in open recirculating water systems.

CASE 16300

| | Values | |
Parameter	City	Softened
pH	7.7	7.8
Alkalinity, ppm as CaCO$_3$	215.0	225.0
Total hardness, ppm as CaCO$_3$	705.0	0.0
Silica, ppm as SiO$_2$	7.0	7.0
Chloride, ppm as NaCl	18.0	78.0*
Total dissolved solids (TDS) by conductance, ppm	607.0	819.0

* The apparent discrepancy in the value of salt, sodium chloride (NaCl), in the city supply is not an analytical error. The author brought this to the attention of the water softener mechanic and was informed that since the softener was delivering soft water, it was functioning satisfactorily. Rebuilding the valves had no effect on the results, nor was the seller of any help. The author acknowledges that something was wrong, but the owners expressed no concern.

It is evident that the limiting factor in the use of the city water is the hardness, or 1200/705 = 1.7 cycles. Should the concentration of the water in the recirculating system become more than 1.7 times that of city water, then scale, or lime, on the heat-exchanger surfaces will be a certainty! Alkalinity (500/215 = 2.3), though high, and silica (150/7 = 21.4) do not enter as factors unless their maximums are exceeded.

Let us now consider the ramifications of operating a centrifugal machine of 100-ton cooling capacity with a minimal chemical feed and bleed-off system; see Fig. 1. To achieve a scale-free operation for the condenser, we must provide a bleed-off rate in excess of the maximum system load (see Fig. 42). Thus, a continuous bleed-off rate of 4.5 gallons per minute (gal/min) must be maintained. We cannot assume a bleed-off rate of 2.2 gal/min, a 50 percent load, adequate with this supply; we must provide for bleed at the maximum anticipated load! At a continuous bleed of 4.5 gal/min, we drain 6480 gal of water as bleed and evaporate, at an actual 100 percent load, 4320 gal for a total daily consumption of 10,800 gal of water!

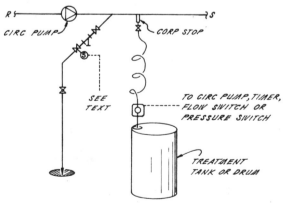

FIGURE 1 Simple chemical feed system.

If soft water were used as make-up water and provisions were made for alkalinity reduction and acid feed, one could reduce water consumption by operating at higher cycles, or so one could be led to believe. Referring to Langelier's saturation index (often abbreviated as LSI, or simply SI) in Fig. 2, let us look at two approaches — 10 cycles with no bleed-off and 7 cycles with a bleed-off of 0.5 gal/min.

10 Cycles of Concentration

Let us add to the known facts that we have a mechanical draft cooling tower with 0.2 percent windage drift and a recirculation rate of 300 gal/min for the 100-ton system. At 10 cycles we would require a bleed-off rate of 0.33 gal/min to maintain the stated concentration of solids in the system. However, since the cooling tower has a built-in bleed-off rate of 0.6 gal/min (the 0.2 percent windage drift), bleed is not required, or so our calculations would lead us to believe. However, let us play it safe and maintain a 0.33-gal/min bleed rate and look at the tower water at 10 cycles:

pH	Being kept at 7.2 to 7.4
Alkalinity	Being kept at 40 to 45 ppm as $CaCO_3$
Hardness	Softener keeping it at 1.0 ppm so we concentrate to 10.0 ppm
Silica	Climbs to 70 ppm
Chlorides	Climbs to 180 or 780 ppm*
TDS	Climbs to 8190 ppm
Sulfates as Na_2SO_4	3122 ppm due to acid use
Langelier's saturation index	$-1.60 I_s$ at 7.4 pH at 140°F (60°C) skin temperature

From the above, clearly at 10 cycles we would wind up with a very corrosive water, $-1.60 I_s$, containing sufficient chlorides and sulfates to challenge any corrosion inhibitor. Furthermore, the use of bio-degradable organic corrosion-inhibitor systems along with the sulfates could lead to problems with bacterial slime or the growth of acid-producing bacterial species. We could hope to control them with algaecides, but would soon be plagued with a strain resistant to the conventional attack and then chlorine, at 50 to 100 ppm, would have to be used.

7 Cycles of Concentration

At this 7 cycles of concentration we would require a bleed-off rate of 0.5 gal/min to maintain conditions as stated. The water, at this level, will appear as follows:

pH	Being kept at 7.2 to 7.4
Alkalinity	Being kept at 40 to 45 ppm as $CaCO_3$

---

* Theoretically, the chlorides should be 180 ppm, but they will be 780 ppm as long as the water softener maintains a 78-ppm constant feed.

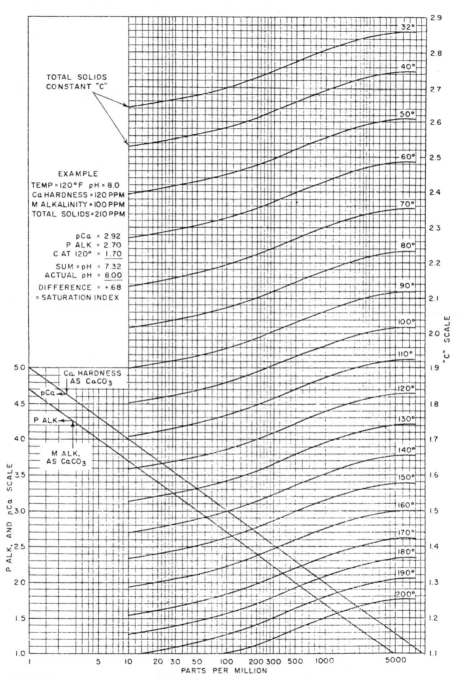

FIGURE 2 Langelier's saturation index I_s. (From Sheppard T. Powell, *Water Conditioning for Industry,* McGraw-Hill Book Company, New York, 1954. Used courtesy McGraw-Hill Book Company.)

Hardness	Concentrates to 7.0 ppm
Silica	Climbs to 49 ppm
Chlorides	126 or 546 ppm as NaCl
TDS	5733 ppm
Sulfates	2172 ppm as Na_2SO_4
LSI	$-1.31/_s$ at 7.4 pH at 140°F (60°C) skin temperature

Even at 7 cycles, clearly there is not too much hope for using this supply as the cooling tower feed. The difference between 10 and 7 cycles is not significant enough to even consider using a water softener. Even as a stop-gap measure, using a water softener would be throwing good money after bad. We would require a softener with a capacity of 200,000 grains per day (gr/day), which would consume about 60 pounds (lb) of salt every time it went into regeneration!

What, then, are the alternatives with this supply? On a contemplated job, the choice should be easy — specify an air-cooled condenser! For an existing job, consider going to a once-through water-cooled system if the switch to an air-cooled condenser is too expensive. With a once-through cooling system, we could consider a threshold-type chemical feed program with, perhaps, a little acid as insurance against scale formation. Figure 3 shows a suggested installation similar to one the author designed for a plant utilizing water on a once-through basis. The baffled mixing tank is constructed in the field out of 12-inch (in) pipe. Note that this is not an installation for a modulated system. When water is required for cooling purposes it should flow at full pipe capacity, and should cease flowing when water for cooling is no longer required.

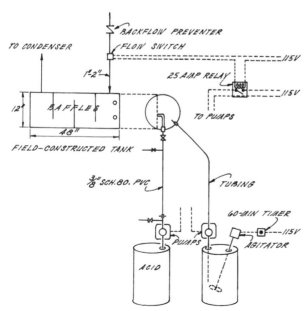

FIGURE 3 Chemical feed for once-through cooling water.

So much for the impossible situation. Now let us consider an acceptable water supply.

CASE 28585

	Values	
Parameter	City	Softened
pH	7.0	7.0
Alkalinity, ppm as $CaCO_3$	90.0	95.0
Total hardness, ppm as $CaCO_3$	125.0	0.0
Silica, ppm as SiO_2	3.0	3.0
Chlorides, ppm as NaCl	17.0	17.0
TDS by conductance, ppm	185.0	132.0

Here the limiting factor is the alkalinity (500/90 = 5.6 cycles) with hardness close behind (1200/125 = 9.6) and silica not even a factor. This supply is satisfactory for an open recirculating water system, but it does pose somewhat of an equipment selection problem for the design engineer at the planning stage. Even though this supply is satisfactory, it will pose a problem because of the alkalinity. Although operation at 5.6 cycles appears to be indicated without acid, prudence dictates otherwise, for a 4.0-cycle maximum without acid would provide a small safety margin. With acid use we could reduce water consumption and operate at 7 cycles with little fear of scale formation.

Let us consider the following, which should assist in the choice of feeding equipment. Use $1 per pound for the scale and corrosion inhibitor, which will be fed at 100 ppm, and $0.08 per pound for 60° Baumé sulfuric acid. The tonnage will be the same as previously — 100-ton centrifugal machine at 100 percent load at 24 h/day at 182 days:

Parameter		
Cycles	4.0	7.0
Bleed-off, gal/min	1.00	0.50
Total bleed, gal/day	1440	724
Treatment (scale/corrosion inhibitor), lb	1.2	0.6
Sulfuric acid, lb	0	2.6
Daily cost	$1.20	$0.81
✕ 182 days	$218.40	$147.42
+ Water at $0.90 per 1000 gal	$236.04	$118.59
Total	$454.44	$266.01

As can be seen, there is a slight benefit in the use of acid for alkalinity control — a 48 percent savings in treatment cost and a reduction in water use of 716 gal/day. Is it cost-effective to sell the client on an acid feed system, given the above figures? The author would be hard-pressed to do so, considering the added cost of the acid feed pump and tank, about $600. However, as insurance against scale formation [a layer of scale the thickness of a dollar bill (0.060 in) increases energy consumption by 53 percent[10]], the use of acid is advised with this supply. The following should help to clarify the point:

100-ton centrifugal machine $\times$ 1.0 kW/ton $\times$ 24 h

$$\times \text{ 182 days} \times \text{\$0.04 per kWh} = \text{\$17,472.00}$$

$$\text{Scale layer of 0.06 in} = \text{\$26,732.00}$$

$$\text{Net waste} = \text{\$ 9,260.00}$$

The above is not a complete case history. It is only meant to show that other factors need to be taken into consideration when you attempt to justify specific chemical feed equipment. On the surface, case 28585 does not show a definite need for acid. But as insurance, the added one-time cost is insignificant compared with the potential added energy costs resulting from scale formation on the heat exchanger.

The following case involves such an excellent supply that it can almost be used without bleed-off.

CASE 47348

Parameter	Value	Cycles
pH	7.0	
Alkalinity, ppm as $CaCO_3$	30.0	500/30 = 16.7
Total hardness, ppm as $CaCO_3$	42.0	1200/42 = 28.6
Silica, ppm as SiO_2	2.7	150/2.7 = 55.6
Chlorides, ppm as NaCl	2.9	
TDS (by conductance), ppm	55.0	

This is an excellent supply for an open recirculating water system. A calculation of the limiting factors reveals that we can operate a spray pond or an atmospheric tower with little, if any, bleed-off at 15 cycles. The following tabulation shows the requirements for each type of heat rejector:

100-ton Refrigeration at 100 Percent Load

	Heat rejector			
	Spray pond	Atmospheric tower	Mechanical draft tower	Evaporative condenser
% Windage drift	1.0	0.3	0.1	0.05
Windage drift, gal/min	1.0	0.3	0.1	0.05
Bleed-off,* 15 cycles	0.2	0.2	0.2	0.2
25 cycles	0.13	0.13	0.13	0.13

* The bleed-off, calculated from $B = E/C - 1$, where C = cycles and E = evaporation, is taken as 3 percent of tonnage in gallons per minute. The use of $E_{gal/min}$ = tonnage $\times$ heat-rejection factor $\times$ 24/(1050 Btu/lb) will provide a much more accurate figure.

The mechanical draft tower and evaporative condenser, owing to their minimal wind-age drift, would require additional water loss, bleed-off, to maintain 15 cycles. With a little acid to neutralize alkalinity, we could conceivably operate at 25 cycles. At this range the only heat rejector that would require bleed-off would be the evaporative condenser, since its windage drift is only 0.05 gal/min.

How important is this ability to read an analysis? Let us take the following case and analyze our options:

Known: A 3000-ton refrigeration load with a mechanical draft tower operating a 100 percent load at 24 h/day. Use 0.03 gal/(min·ton) as an evaporation factor.

Limit: The authorities will not allow more than 1000 gal/day to be dumped from the system.

Options:

	Cycles		
	10	15	25
Wind drift, gal/min	3.0	3.0	3.0
Bleed-off, gal/min	10.0	6.43	3.75
Net loss, gal/min	7.0	3.43	0.75
Net loss, gal/day	10,080	4939	1080
Cost/day*			
Scale/corrosion inhibitor	12.00	8.00	4.75
Acid	0	0	3.17

* Inhibitor is fed at 100 ppm at $1.00 per pound with acid at $0.08 per pound.

A casual inspection of the water analysis, case 47348, would lead us to believe that acid is not required. However, when confronted with the 1000-gal/day maximum waste limitation, do we inform the client that the project is not feasible? An in-depth analysis would reveal that by going to 25 cycles and using an easily maintained sand filtration system, a water-meter-activated chemical and acid feed system with a tower capable of a slightly higher windage drift, we could satisfy the limiting criteria.

So far we have covered an impossible supply, an acceptable supply, and an excellent supply, all of which had one thing in common — *consistency!* Let us now examine and address the problem of the variable supply, the scourge of water treatment firms and thorn in the side of design engineers.

Figure 4 lists analytical results of a central New York municipal water supply covering the 1981 to 1983 air conditioning season. How can one propose to treat this water in an effective manner if one is not aware of the nature of the supply? If one were to accept the results of a springtime analysis as representative of the supply, for example, one dated April 20, 1982, one would be hard-pressed to explain to the operators why the heat

FIGURE 4 Analytical results of domestic water, city of Ithaca, NY, 1981, 1982, 1983

Date 1981	5/11	5/18	5/26	6/1	6/9	6/15	6/16	6/22	6/29	7/6	7/13	8/17
pH	7.5	7.5	7.6	7.3	7.2	7.5	7.4	7.1	7.4	7.4	7.0	7.2
M alkalinity, ppm as $CaCO_3$	75	85	80	90	85	90	105	100	115	125	100	110
Hardness, ppm as $CaCO_3$	98	104	104	114	112	114	128	120	140	136	124	130
Silica, ppm as SiO_2	4.0	4.0	3.2	4.0	3.4	3.0	5.0	4.4	6.0	6.0	5.3	7.0
Sodium chloride, ppm	41	35	35	35	29	32	38	35	52	46	35	41
TDS by meter, ppm	154	154	159	121	148	150	149	154	174	180	167	169

Date 1982	4/20	5/7	5/13	5/14	5/19	5/24	6/2	9/22	9/28	10/5	10/13
pH	7.4	7.4	7.7	7.3	7.4	7.2	7.6	7.4	7.6	7.6	(6.8)
M alkalinity, ppm as $CaCO_3$	60	95	100	80	95	110	115	130	140	130	135
Hardness, ppm as $CaCO_3$	90	102	120	112	130	142	142	170	152	184	188
Silica, ppm as SiO_2	5.0	3.8	5.0	· · ·	4.2	(25)	4.0	5.0	5.0	5.0	4.0
Sodium chloride, ppm	35	35	29	41	35	41	35	47	41	35	41
TDS by meter, ppm	113	139	173	132	145	187	196	197	203	185	202

Date 1983	5/2	5/9	5/20	5/30	6/10	6/21	6/30	7/6	7/11	7/29	8/12
pH	(8.4)	(8.3)	7.4	7.1	7.4	7.7	7.8	7.6	7.8	7.5	7.2
M alkalinity, ppm as $CaCO_3$	55	70	70	90	105	110	110	115	115	125	125
Hardness, ppm as $CaCO_3$	70	100	120	108	275	144	130	140	135	154	148
Silica, ppm as SiO_2	4.0	(40.0)¡	14.	4.1	3.1	2.0	4.0	5.0	4.6	5.0	4.9
Sodium chloride, ppm	17	23	23	35	29	41	29	41	41	26	47
TDS by meter, ppm	103	124	134	167	162	164	178	197	179	191	185

NOTE: Figures in parentheses indicate possible analytical error. All figures are for the city of Ithaca, NY.

rejector or exchanger was loaded with lime at season's end. If one had based one's recommendations on a summer analysis, say, one dated August 17, 1981, and supplied an acid-based treatment, then the pH would have taken a nose dive during the spring months. Figure 74 is a graphical representation of the alkalinity during the period of observation, and it illustrates the difficulties involved in setting up a proper water treatment program. Thus the reason for stressing that when a set of water treatment specifications is put together, the onus of responsibility should fall on the water treatment firm. The hardness was not graphed because it does not have as significant an impact as does the change in alkalinity.

How would we attack the problem of a variable water supply? First, and foremost, we obtain a history of the water. Hopefully the local water authorities can supply it; if not, then a few calls to others using the supply can produce analyses provided by their water treatment firms. For this particular supply (Fig. 4), the author would insist on nothing less than a water-meter-activated chemical feed system, a pH controller/recorder for acid feed, and the maintenance of the concentration at 6 cycles, not higher! With small systems under 100 tons, to operate at higher than 3.0 cycles with a simple chemical feed/bleed system (Fig. 1) is to risk scale formation and an inefficient operation.

At what tonnage does one insist on a pH controller? The design engineer must make this decision based on payback, cost of treatment chemicals, and cost and availability of the water supply. Figure 5 provides some guideline data. In each case, the difference between using acid and not using acid is 68 percent. However, with a cost factor of $1000 for an acid pump and a pH controller, the payback is 6 years for the 200-ton unit and 4 years for the 300-ton unit. The reader should, at this time, study the contents of pages 6 and 7 as they relate to costs vs. scale, for the numbers can be misleading. To decide, based on a 5-year payback, not to use acid could be a mistake. Once a unit develops scale, it can cost anywhere between $600 and $1000 to remove. This does not take into consideration three factors:

1. Increased energy costs

2. Damage to fan or air-moving components of system[11]

3. Stripping of the galvanizing if an improper acid is used for descaling a galvanized tower or evaporative condenser

There are many facets to water treatment for open recirculating water systems. It involves more than a couple of drums of chemical, more than feeding equipment, more than a choice of water treatment firm. And it is folly to concentrate on one aspect of the system and overlook the impact of the whole.

FIGURE 5 Cost comparisons of various tonnages with and without acid use in treatment program

Operation: 100 percent load @ 24 h/day @ 125 days
Treatment: Inhibitor @ 100 ppm @ $1.00/lb, acid @ $0.08/lb

Tonnage:	100		200		300		400		500	
Cycles	3	6	3	6	3	6	3	6	3	6
Bleed-off, gal/min	1.5	0.6	3.0	1.2	4.5	1.8	6.0	2.4	7.5	3.0
Treatment, lb	1.75	0.75	3.5	1.5	5.5	2.25	7.0	3.0	9.0	3.5
Acid, lb		3.6		7.3		11.0		15.0		18.0
Daily cost*	1.75	1.04	3.50	2.08	5.50	3.13	7.00	4.20	9.00	4.94
Season	219	130	438	260	688	391	875	525	1125	618
Difference	$89.00		$178.00		$297.00		$350.00		$507.00	

* The daily cost does not reflect costs due to increased water consumption (bleed-off) or costs involved in control of organic growths.

11

DOMESTIC WATER TREATMENT

When to provide treatment for a domestic water supply is, unfortunately, decided after a problem has manifested itself or as local experience dictates it (for example, all users of the supply have installed water softeners, so one must be specified). This approach, although workable, is not good because of the nature of water. The reader, the user, must understand that water, as supplied by municipalities, is not that simple H_2O molecule depicted in elementary chemistry textbooks, but rather a complex soup of many ingredients. Each water supply is distinct and different from every other, so what works for one may not work for another.

If one is directed to deal with a water supply problem, one can study the situation and specify the necessary equipment to correct it. However, most specifications involving domestic water systems call for the addition of a water softener when the water is hard enough to warrant it, but not because a study of the supply at a particular temperature dictated it. It is the objective of this section to provide the reader with a rational way to address domestic water problems.

The primary analytical tool to use in choosing water treatment is the *water analysis*. Let us consider the four previous cases and make extensive predictive use of Langelier's saturation index I_s (Fig. 2).[1] Here I_s is a predictive tool and so should not be taken as an absolute. Clearly other factors contribute to problems, e.g., dissolved gases, chlorine, chlorides, sulfates, particulates, water velocity, etc., and must be taken into consideration.

CASE 16300

Parameter	Values		
	Raw	Soft	Theoretical
pH	7.7	7.8	7.7
Alkalinity, ppm	215	225	215
Hardness, ppm	705	0.0	1.0
Chlorides, ppm	18	78	18
TDS by conductance, ppm	607	819	607

Figure 6A, B, and C lists the saturation index for this supply from 40 to 180°F in 10-degree increments. Figure 7, curves A, B, and C, shows plots of the saturation index to illustrate the trend in change of temperature. The raw supply, curve A, starts off as a mild scale former at 40°F, with $I_s = +0.5$, and worsens as the temperature climbs. This explains why scale forms on the hottest parts of a hot-water heater, heat exchanger, boiler tubes, or second pass of an absorption machine. To soften the supply (curves B and C), a common cure-all, changes the nature of the water and makes it very corrosive.

Although 100 percent softening is common, it should not be thought of as the answer, unless soft water is required for a specific purpose, e.g., boiler make-up water. Soft water, being corrosive, will attack ferrous and nonferrous metals in the system and thus add lead, tin, copper, and other undesirable metals to the supply.

Let us now address the question of how to treat this water supply. For drinking purposes, it should not be softened even though lime may be causing plumbing problems. Softening creates two perils of immediate concern:

1. The softener converts all the lime, $CaCO_3$, to soda ash, Na_2CO_3, or sodium carbonate, which is detrimental to those on low-sodium diets.

2. In the conversion from a hard to a soft water supply, the water becomes corrosive, so it will attack galvanized-steel and copper pipes as well as leach out solder (lead) from joints. The amount leached out may not be sufficient to cause concern; but, then again, it may.

A good approach would be to treat the cold-water supply with a food-grade glassy polyphosphate to sequester, or hold in, the hardness. At a dosage rate of 3.0 ppm (3 lb per 119,832 gal of water), we would be adding a negligible amount of sodium to the supply; in the softening process, however, an amount equal to the total hardness plus that contributed by a malfunctioning softener valve would be added. The objective is to hold the lime in solution, since it wants to come out of solution as the temperature increases. Figure 7A shows this occurring as the curve climbs. If the temperature is kept very constant, then adding a sequestrant lets us operate with reduced probability of problems.

In using polyphosphates, the author recommends no more treatment be mixed in the treatment tank than can be consumed within 4 to 5 working days. Polyphosphates, in time, have a tendency to change to the orthophosphate form, and thus:

1. Will not provide the desired scale protection

2. Will react with calcium, lime, in the water to form a calcium phosphate sludge which will plug lines and impede heat transfer if it builds up on the heat-exchanger (HX) surfaces

3. Could provide a breeding ground for unicellular life forms, algae, etc., which could affect the water's potability

To forestall problems, flush the treatment tank, pump, and lines on a periodic basis with hot water, at 180°F; in addition, there should be a final rinse with chlorine while the feed line is disconnected from the system. Tests for TDS and orthophosphate should be performed no less often than weekly and adjustments made accordingly. Daily tests are the ideal, but management balks at the idea since testing is not considered to be a productive pastime for an employee. As for sending samples to a laboratory, there is an appeal to this approach — the client believes something is being provided, and the chemical supplier is given the opportunity to be of service. However, sending samples for testing should be done only to augment the program, not in lieu of it. In sending samples, one is at the mercy of time.

What about the domestic hot-water supply? There is no doubt that this supply should be softened, but to what extent? An examination of Fig. 7B and C shows what happens to this supply once it is softened. For both curves I_s is less than -1.0, which makes the

FIGURE 6 Saturation index of three water supplies

°F

CASE 16300		40	50	60	70	80	90	100	110	120	130	140	150	160	170	180
		Actual Hard Water Supply, A														
pH = 7.7	pCa	2.15														
Alk. = 215	pAlk	2.38														
Hard. = 705	C	2.67	2.53	2.41	2.28	2.15	2.04	1.94	1.85	1.75	1.67	1.58	1.51	1.43	1.35	1.29
TDS = 607	pH_s	7.2	7.06	6.94	6.91	6.68	6.57	6.47	6.38	6.28	6.20	6.11	6.04	5.96	5.88	5.82
NaCl = 18	I_s	+0.5	0.64	0.76	0.89	1.02	1.13	1.23	1.32	1.42	1.50	1.59	1.66	1.74	1.82	1.88
		Theoretical Soft Water Supply from above, B														
pH = 7.7	pCa	5.00														
Alk. = 215	pAlk	2.38														
Hard. = 1.0	C	2.67	2.53	2.41	2.28	2.15	2.04	1.94	1.85	1.75	1.67	1.58	1.51	1.43	1.35	1.29
TDS = 607	pH_s	10.05	9.91	9.79	9.66	9.53	9.42	9.32	9.23	9.13	9.05	8.96	8.89	8.81	8.73	8.67
NaCl = 18	I_s	−2.35	2.21	2.09	1.96	1.83	1.72	1.62	1.53	1.43	1.35	1.26	1.19	1.11	1.03	0.97
		Actual Soft Water Supply from Case 16300, C														
pH = 7.8	pCa	5.00														
Alk. = 225	pAlk	2.33														
Hard. = 0.0	C	2.70	2.53	2.42	2.29	2.17	2.05	1.95	1.87	1.77	1.66	1.59	1.52	1.44	1.36	1.31
TDS = 918	pH_s	10.03	9.86	9.75	9.62	9.50	9.38	9.28	9.20	9.10	8.99	8.92	8.85	8.77	8.69	8.64
NaCl = 78	I_s	−2.23	2.06	1.95	1.82	1.70	1.58	1.48	1.40	1.30	1.19	1.12	1.05	0.97	0.89	0.84

°F

CASE 28585		40	110	180
		Actual Supply, D		
pH = 7.4	pCa	2.90		

14

Alk. = 90				
Hard. = 125				
TDS = 185				
pAlk	2.75			
C	2.61	1.79	1.25	
pH$_s$	8.26	7.44	6.90	
I$_s$	−0.86	−0.04	+0.50	

Actual Supply, E

pH = 7.4	pCa	5.00		
Alk. = 95	pAlk	2.73		
Hard. = 0.0	C	2.59	1.78	1.22
TDS = 132	pH$_s$	10.32	9.51	8.95
	I$_s$	−2.92	2.11	1.55

CASE 47348		°F			
		40	110	180	190

Actual Supply, F

pH = 7.0	pCa	3.37			
Alk. = 30	pAlk	3.22			
Hard. = 42	C	2.57	1.75	1.19	1.13
TDS = 55	pH$_s$	9.16	8.34	7.78	7.72
	I$_s$	−2.16	1.34	0.78	0.72

Theoretical Supply, G

pH = 7.0	pCa	5.00			
Alk. = 30	pAlk	3.22			
Hard. = 0.0	C	2.57	1.75	1.75	1.19
TDS = 55	pH$_s$	10.79	9.97	9.41	
	I$_s$	−3.79	2.97	2.41	

supply quite corrosive under 140°F. At 180°F, $I_s = -0.8$, still quite corrosive. The solution to this problem is *partial softening*. It makes no sense to do 100 percent softening. The objective is to provide a supply that is not scale-forming, not scum-forming, and noncorrosive. So to accomplish our goal, we mix the hard water and 100 percent soft water to obtain a mixed supply meeting the objective. In this case, we need to obtain a final mixture with 4.0-ppm hardness. At this range, we would have a supply with a saturation index quite close to zero but still very slightly negative at 140°F. If the temperature were 180°F, then we would determine the correct mix at that temperature needed to achieve the stated goals.

Let us now consider the moderate supply, case 28585.

CASE 28585

Parameter	Values	
	Raw	Soft (actual)
pH	7.4	7.4
Alkalinity, ppm	90.0	95.0
Hardness, ppm	125	0.0
Chlorides, ppm	17.0	17.0
TDS by conductance, ppm	185.0	132.0

This supply is considered to be scale-forming because its hardness is 125 ppm (or 7.3 gr/gal). Curve *D* in Fig. 7 shows this to be true at temperatures above 140°F. At temperatures above 180°F, the sanitizing temperature, we can expect problems with HXs supplying dishwashers and similar equipment. Odd as it may seem, this water supply is also quite corrosive at the cold end, 40 to 50°F, since the I_s is below zero at those temperatures. As a cold-water supply, we can expect it to corrode the cold-water distribution system. The blue (copper) and brown (iron) stains below dripping faucets indicate this problem. Remember that ingestion of these corrosion products — copper, lead, tin, and zinc — is possible where the supply is corrosive. If there is sufficient reason for concern, this cold-water supply could be treated with a coating material, e.g., sodium silicate or water glass, to minimize the corrosive action of the supply.

Since this supply could cause problems at 180°F, lime on HXs and dirty glasswear, we might make a recommendation to soften the water. An analysis of the softened-water, the discussion of supply *E* in Fig. 6, shows that we have traded a mildly scale-forming supply, $I_s = +0.50$ at 180°F (supply *D*), for a very aggressive one, $I_s = -1.55$ at 180°F. Is 100 percent softening the answer? Again we ask, What are we trying to do? Let us examine each system and choose the alternatives to meet our objectives.

Cold-Water Supply

The water has a metallic taste, and the sinks are stained with brown and blue material. The problem has been determined to be corrosion, and the objective is to fix the water so it does not stain the fixtures.

Since $I_s = -0.86$ at 40°F and since the pH is 7.4, we should not concern ourselves

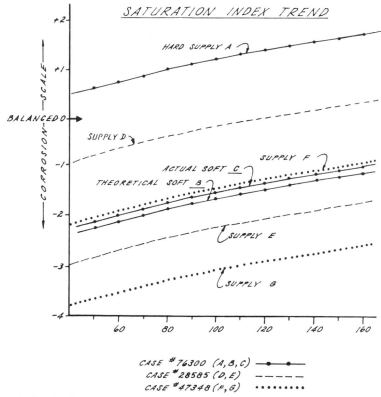

FIGURE 7 Saturation index trend.

with neutralizing chemicals. Since the stains are brown (iron rust) and blue (copper), we have a bimetallic situation. The use of polyphosphates will control the problem of iron deposits but may not stop the copper deposits. An analysis of the contributors to the I_s (see Fig. 6) tells us that we should do whatever we can to get the pH (7.4) and pH_s (8.26) to coincide, since $I_s = pH - pH_s$, and thus bring I_s close to zero. By increasing the hardness (sounds a bit out of the ordinary to add hardness to a hard-water supply in order to control corrosion, does it not?), we cause I_s to approach zero. We can also achieve our objective by increasing the pH and hardness by adding slaked lime, $Ca(OH)_2$, to the supply. Another approach would involve adding sodium silicate (water glass) at the correct pH level to increase the pH and then to coat the internal parts of the system with a layer of glass. Since we want to stop the staining problem, we can take any route that will make the cold water (40 to 50°F) less corrosive.

Hot-Water Supply

Here the problem is scale formation on the HX tubes, so it should be straightforward to fix it, right? Again, first we must analyze the situation and determine all the parameters involved. If we try a hard/soft mixture at 140°F, we will still have a problem at 180°F, the sanitizing temperature, even though I_s may be only about $+0.3$, or very slightly scale-forming, and so cause a mild staining of glassware. If we go to 100 percent softening (curve *E*, Fig. 7, and discussion of Fig. 6) and maintain temperatures close to 140°F, we will have a supply with an I_s of -1.85 which will cause problems since this is quite corrosive. At temperatures near 180°F, we will have a supply with an I_s of -1.60 which, while better than the results at 140°F, is still just as corrosive.

The solution for this supply is to mix hard and soft water for the particular temperature at which a problem is being experienced. We must take into consideration the fact that the calculations should be made at the skin temperature of the HX unit. That is, we may want water at 120°F, but we should calculate I_s at a slightly higher temperature since scale formation occurs at the higher skin temperature of the HX. Let us decide that we will supply 160°F water. So we calculate for 180°F water and try to achieve a supply with a pH of 7.4, a 35-ppm hardness, an alkalinity of 90 ppm, and a TDS of 185 ppm. This should require a mixture of 70 percent soft water and 30 percent hard water. Calculating I_s at 180°F gives us an index of -0.04, very close to neutral, so we should not be experiencing scale formation at the HX tubes; and at 160°F the index is -0.15.

To the final mixed supply we add a little polyphosphate as insurance against corrosion, 3 to 5 ppm as Na_3PO_4 (trisodium phosphate). The final supply should be just very slightly scale-forming or corrosive, with sufficient inhibitors to protect against scale formation and/or corrosion.

Let us now consider the soft-water supply, case 47348.

CASE 47348

Parameter	Values	
	Raw	Soft (theoretical)
pH	7.0	7.0
Alkalinity, ppm	30	30
Hardness, ppm	42.0	0.0
Chlorides, ppm	2.9	2.9
TDS by conductance, ppm	55.0	55.0

Here we have a natural soft-water supply, curve *F* in Fig. 7, that is very corrosive to the cold-water distribution system ($I_s = -2.16$ at 40°F). As with previous examples, there is a reduction in corrosivity with an increase in temperature; however, even at 190°F $I_s = -0.72$. A water supply this corrosive would normally be altered by the supplier, if only to protect the distribution system. However, some suppliers may not alter the water, so the end-user should make provisions to correct the corrosive tendency. The best approach, as previously suggested, is to alter the supply according to the intended temperature. The cold-water supply is altered with lime [$Ca(OH)_2$] to bring I_s close to zero. The hot-water supply should be treated separately and provided with its own program.

In Fig. 7, curve *G*, a plot is drawn of the theoretical soft-water supply, to show how corrosive the supply can get. Based on this extrapolation, clearly softening is not a recommended procedure.

The three cases outlined — very hard, hard, and soft water, as commonly defined — were shown to require treatment, not by guesswork but by using the available tools: water analyses and Langelier's saturation index. The water analyses, preferably a series of analyses taken over an extended period, should include the following:

1. pH (using slides or a pH meter)

2. Alkalinity, expressed as calcium carbonate ($CaCO_3$)

3. Total hardness, expressed as $CaCO_3$

4. TDS, preferably by evaporation, but a conductance guesstimate is acceptable

5. Chlorine, expressed as free chlorine

6. Chlorides, expressed as sodium chloride (NaCl)

7. Sulfates, expressed as sodium sulfate (Na_2SO_4)

The chlorine content, although not taken into consideration earlier, is nonetheless important since concentrations of 2.0 ppm are suspected of accelerating the corrosion of copper pipe. Although the author has tried to keep this chapter as simple as possible, the reader should understand that treatment is not a simple, cut-and-dried procedure. To illustrate the difficulties, let us review case 28585, the hard-water supply, as it is used as a domestic, circulating, hot-water supply at 180°F (82.2°C).

The water enters at a temperature of 40°F (4.4°C), pH of 7.4, hardness of 125 ppm, TDS of 125 ppm, and dissolved oxygen level of more than 5.0 ppm. As the temperature is increased to 180°F (82.2°C), the solubility of lime decreases and lime builds up in a nonuniform layer on the HX surfaces. At 180°F (82.2°C), the solubility of calcium carbonate is reduced to 50 ppm, so we are left with a relatively soft water supply owing to scale formation. Recalculating with a pH of 7.4, hardness of 50 ppm, TDS of 110 ppm, and alkalinity of 90 ppm at 180°F (82.2°C) gives us a supply with an index of about +0.13, close to the balanced condition. Now if we take the effect of the dissolved oxygen into consideration, a corrosion rate in excess of 10 mils/yr would not be unrealistic for the ferrous components. Given the effects of particulates circulating in the system, rust, and chunks of lime, we can expect erosion to shorten system life.

With the analytical tools available and the reasons for treating domestic water systems explained, the design engineer is in a position to make definitive recommendations. The recommendations should be not a simplistic approach, but a preventive maintenance approach to a particular or potential problem. The treatment may involve nothing more than controlling system temperature, which could avert scale formation; or it may be as complex as mixing waters and adding chemicals to achieve an end result. Nonetheless, it will be an approach based on sound principles.

STEAM BOILERS

Very little justification is needed in recommending treatment for steam boilers; it is amazing how much information is available and how little of it is put to use. For example, it is reported[12] that chlorides in excess of 100 ppm cause pitting of boiler tubes even when they are protected. But how often is that information taken into consideration when the boilers are being planned? It is the author's hope that the reader will avoid, at the design and planning stage, those problems seen to cause discord among the owners, engineer, mechanical contractor, and water treatment supplier.

Water treatment is a very small item in the construction budget which, if improperly applied as a result of restrictions imposed at the design stage, can cause severe equipment damage. An example is a multiple-boiler, low-pressure, steam-heating system where the boilers are interconnected to prevent flooding and thus operational problems (Fig. 8).

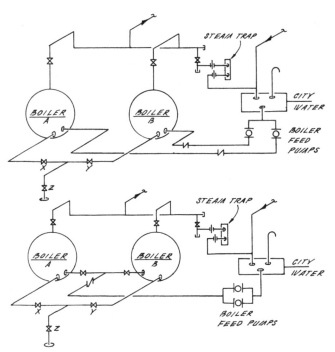

FIGURE 8 Typical low-pressure steam-heating boilers.

In normal operation, one boiler is operated and the other is left on standby, idle. Under this constraint, when the header valves are open, the operating boiler will attempt to flood the idle unit with condensate. However, since there is an equalizing set-up (valves X and Y are open with Z closed), the idle unit will not flood but fill to the operating level with condensate, with the rationale being as follows: Steam from the on unit will enter the idle unit via the open header valves and condense, since the idle unit, by virtue of its being off, is cooler. As the condensation increases, it tends to push the treated boiler water to the operating unit. The reader should also expect thermal migration to affect this water movement.

The author suspects there is a twofold belief that

1. Treated boiler water is condensing in the idle unit.

2. The idle unit is protected since it was initially treated.

However, field experience has proved this not to be the case. Condensate, with a relative density of about 0.98 at $70\,°C$ ($158\,°F$), slowly builds up on top of the treated boiler water, which has a higher relative density, greater than 1.00. Since the water in the idle unit is not being agitated, there will be a distinct separation between the layers of water, with the lighter water on top. As condensate continues to enter the idle unit, all the treatment and dissolved solids will be pushed into the operating unit. The consequences of this interaction are as follows:

1. The idle unit, on filling with condensate, will corrode, unless it is made of cast iron.

2. The operating boiler will fill with dissolved solids, its own complement plus that contributed by idle unit, which could lead to:

    a. Priming/foaming due to high level of solids.

    b. Scale buildup.

    c. Attack on glass and bronze components owing to alkalinity buildup.

The solution to the problem has been, and probably continues to be, as follows:

1. Alternate the boilers on a scheduled basis, activated either manually or by a timer.

2. Valve the boilers off and hope the valves hold.

3. Have both boilers fire simultaneously from one set of controls.

Experience has shown that these approaches are far from solutions to the boiler-flooding problem and create their own difficulties. The first approach assumes the load to be continuously uniform, the second assumes that the valves will hold, and the third allows one control to determine whether the boilers will operate at all. Certainly all are unacceptable!

The author has suggested a solution[13] that has been used at one location with a measure of success — the installation of trapped header swing check valves, as shown in Fig. 9. Here, the header valves are left open since the swing check valves will control the flow of steam from each boiler according to demand. The steam traps are required to keep the valve body free of trapped condensate, which could affect operation of the swing check valve.

The swing-check-valve system operates as follows: With an idle boiler, the pressure of the on unit will keep the check valve shut. The little condensate that builds up behind the valve is bled away by the traps. If the load is insufficient for one boiler, then the idle unit comes on and opens the check valve as the pressure from it rises above the pressure from the other unit holding the check valve shut. After the original work was done, the author found vacuum breakers to be a necessity. Otherwise, the suction created as a boiler shuts down will draw water from the condensate feed tank.

Because of equipment and installation costs, the swing-check-valve system has not been too popular, although fear of hammer may also be a deciding factor. The author continued to search for a better solution to boiler flooding and discovered an excellent approach[14] which is less expensive, in terms of equipment cost and installation, does not have potential for hammer, and works! Figure 10 shows the suggested solution.

Here each boiler has a steam trap to draw off excess condensate from the internal parts. The header valves are left open, and condensate is allowed to build up. As soon as the liquid level reaches the trap pipe, about 1 in

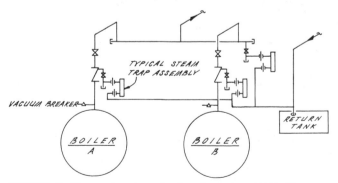

FIGURE 9 Trapped swing check valve — low-pressure steam-heating boilers.

above the high-water mark, the boiler pressure pushes the condensate out of the boiler. Note that the 1-in figure is not fixed or magical — do not hesitate to experiment. However, the higher you raise the pipe, the greater the amount of condensate that can build up in the boiler, which must be compensated by an equal amount of make-up water. If the amount of condensate allowed to remain in the boiler is greater than the capacity of the storage tank, the return tank will flood when the unit is called on to produce steam.

Both suggested installations (Fig. 10) have the threaded, or welded, pipe about 1 in above the boiler's high-water level. In boiler A it comes in through the top, and in boiler B it comes through the side via the fitting where the skimmer blowdown plug is located. Boiler A depends on system pressure to push the condensate up the pipe,* and in B it just flows out of the system. Both drawings are of actual installations on Cleaver Brooks boilers operating at 15 pounds per square inch gauge (psig), or less. There is no reason to believe the method will not work with other boilers as long as the fittings are there, i.e., as a means for installing the trap. The cast-iron sectional boilers pose the biggest challenge, but they can be piped similarly.

Let us now consider the water analysis, our primary tool, and determine the options available for a steam boiler operation. However, let us do so with two important terms clearly defined: *heating* and *process*.

Heating: A single- or multiple-boiler installation where the steam is used to transfer heat which, on condensing, is brought back to the condensate return tank. The condensate loss is not to exceed 3 percent per day without steps being taken to reduce the loss.

Process: A single- or multiple-boiler installation where the steam is used in such a manner as to preclude its return to the condensate tank. The loss will vary depending on the process load.

---

* The author has been informed that it is suction that draws up the condensate; however, the explanation is not altogether satisfactory.

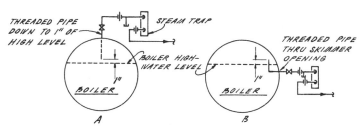

FIGURE 10   Surface-trapped low-pressure steam-heating boilers.

Obviously pressure is a factor in the above operations; however, in this chapter we concern ourselves with units operating below 300 psig. Let us now analyze each case as in previous sections and determine available options with the various supplies for the following unit: two 100-horse-power (hp) fire-tube boilers, 3450 lb of steam per hour (or 413 gal/h) per unit at 100 percent load at 24 h/day with no flooding problems and steam traps operating as designed.

CASE 54937

Parameter	Values
pH	7.7
Alkalinity, ppm as $CaCO_3$	180.0
Total hardness, ppm as $CaCO_3$	264.0
Silica, ppm as SiO_2	10.9
Chlorides, ppm as NaCl	233.0
TDS by conductance, ppm	485.0

Heating

There is no doubt that the water in this supply must be softened prior to being used in a steam boiler for heating purposes. However, even with softening we are faced with significant potential problems. At 2 cycles of concentration, which is not unusual during the first months of operation, we would be faced with two problems:

1. A carbon dioxide (CO_2) content of 232 ppm, which translates to 327 ppm of carbonic acid (H_2CO_3) to attack return lines

2. A doubling of chlorides, to 466 ppm

The author has treated heating boilers in schools by using this softened water supply for more than 12 years with no appreciable problems even though the chloride level, as NaCl, has been in excess of 900 ppm. However, the treatment program—chromates at 2000 to 2500 ppm—has been maintained and the condensate loss minimized. If care had not been exercised and problems had been left uncorrected, conceivably boiler repair costs (e.g., tube replacement, descaling, return-line corrosion, etc.) would have proved to be a heavy financial burden on the school district.

Another approach with this water supply would have been to recommend the installation of a demineralizer. However, at a cost of $6000 to $10,000, the suggestion would have fallen on deaf ears, especially given maintenance costs. And another alternative which could have been considered, especially with the use of thinner boiler tubes, was the utilization of *cistern water*.

The author has had firsthand experience in the use of cistern rain/snow melt water for three 150-hp fire-tube Cleaver Brooks boilers used for heating and humidification, with excellent results! The cistern measured approximately 20 × 20 × 4 ft (1600 ft³) and was able to supply all the water required for heating and humidification. The cistern supply did drop quite low during a dry spell but fully recovered after the first significant precipitation. Blowdown was kept to a bare minimum. The low-water safeties were tested weekly, with minimum water loss, and the boiler was provided with monthly 5-second(s), front and rear, bottom blowdowns. Is the use of cistern water feasible where the water is hard in lieu of a water softener? Let us consider a hypothetical case and weigh our options.

Structure: A small school building with 20,000 ft^2 of roof surface.

Equipment: Two 100-hp fire-tube boilers containing 400 gal of water per unit and operating at 15 psig.

Particulars: Each boiler is equipped with antiflooding steam traps, and all traps are in excellent condition. One boiler is capable of handling 100 percent load during the coldest anticipated period. In normal operation one unit is on lead, operating at 50 percent load at 24 h/day at 125 days, with the other unit on lag, or standby. Condensate loss is minimal — not greater than 13 gal/day during heating season. The treatment program consists of a nitrite/borate formulation that keeps the nitrite, as $NaNO_2$, within 2000 to 2500 ppm.

Options:

1. *Use raw city water, case 54937, as is.*

   Since the treatment has to be maintained at a minimum of 2000 ppm, we cannot allow the city water to contribute in excess of 1500 ppm of solids; otherwise, the boiler could experience operational difficulties, foaming, priming, etc. If the treatment contains suitable chelating (softening) and antifoam agents, we could conceivably operate the boilers at 2 or 3 cycles with a measure of comfort. However, the particulars state that the system could lose up to 13 gal/day of condensate, one volume of boiler water per month, so blowdown and daily chemical tests are a must! Under these conditions we could expect to use about 25 gal of treatment water per season, assuming that condensate loss does not exceed 13 gal/day and the water supply does not change. Even with an ideal treatment program, we can expect some lime since the water is quite hard. Also, we should not overlook the potential for problems resulting from the high sodium chloride content, especially at gaskets and where tubes are rolled in.

2. *Take the water and soften it.*

   The normal route. However, as previously suggested, this approach contributes to return-line corrosion by changing all the lime to soda ash via the following (where R = ion exchange resin):

$$\text{Lime} \quad \text{resin} \qquad\qquad \text{resin} \quad \text{soda ash}$$
$$CaCO_3 + Na_2R \longrightarrow CaR + Na_2CO_3$$
$$\diagdown \ \text{regeneration} \ \diagup$$
$$+ NaCl, \text{ salt}$$

$$\text{Soda ash} \quad \text{water} \qquad \begin{array}{c}\text{carbon}\\ \text{dioxide}\end{array}$$
$$Na_2CO_3 + 2H_2O \xrightarrow{\quad\blacktriangle\quad} CO_2 \uparrow + 2NaOH + H_2O$$

The soda ash in the boiler breaks down and releases carbon dioxide, which forms carbonic acid with condensate to destroy return lines, traps, etc. The boiler water, per se, also becomes quite alkaline, and attack on gauge glass and bronze components is possible.

With the supply water softened, we could operate at higher cycles of concentration — 4 cycles would seem to be maximum with the 2000-ppm nitrite treatment program. At this concentration we would have 2000-ppm nitrite plus 1940-ppm TDS from the water supply plus the borate, which could prove to be too much for the boiler water to hold at that pressure and temperature, resulting in foaming, priming, etc. Also keep in mind that softening does not remove chlorides, and a bad valve could increase the chloride level, so at 4 cycles we should consider the potential negative effects of 932 ppm of sodium chloride.

3. *Use cistern water.*

   With a 20,000-ft^2 roof, each 1 in of rainfall could provide approximately 12,000 gal of water. With a cylindrical tank 5 $\times$ 8 ft, we could store about 1000 gal, or 2 months' reserve at a 13-gal/day use. Surely, 1 in of rain/snow melt each 60 days is not unreasonable to expect. Even a smaller building, 5000 ft^2, could still provide 3000 gal/in of downpour. In lieu of an inside 5 $\times$ 8-ft cylindrical tank, for new construction we could specify a closed concrete cistern 10 $\times$ 10 $\times$ 8 ft, with a plastic liner or bitumismatic coating. This tank could hold 6000 gal of water, one good downpour, and carry the two boilers, at 13-gal/day maximum water loss, for the entire winter with a little extra for a bad trap or two.

As can be seen, the use of cistern water is not only feasible but also highly attractive for all low-pressure steam-heating boilers as long as condensate loss is not a significant factor. Even when the condensate loss is high, as long as it is taken into consideration (e.g., humidification losses) and reasonable projections are made, there should be no hesitation about looking into the use of cistern water. The advantages of using cistern water should be obvious:

1. The water is almost distilled, so solids buildup would be very slow, if it occurred at all.

2. There are no chlorides to worry about, other than those gleaned from the air.

3. The level of dissolved gases would be low, so return-line corrosion, especially from CO_2 attack, would be low.

4. There is no complicated water-softener valve to worry about.

5. Other than the initial installation cost, there are no significant upkeep costs.

6. Since blowdown is kept to bare minimum, treatment loss is quite low.

Figure 11 shows the basics of a cistern supply; however, it should be understood that not all locations are alike and that the mechanics of installation must be feasible. The information is provided only as a guide. The storage tank is only limited by space considerations and cost. The two pumps are sized such that the pump is capable of maintaining adequate water feed should condensate cease to enter the return tank. The pressure tank may or may not be needed, since its purpose is only to maintain a constant pressure on the line feeding the return tank.

Process

For process use, the water in case 54937 must be subjected to treatment. The author is very partial to demineralization, and that is the suggested treatment. The author feels that it is just as important to reduce the chloride content of the supply as it is to reduce the hardness and alkalinity.

We have already seen how softening affects this water supply — the lime ($CaCO_3$) is converted to soda ash (Na_2CO_3) which under heat and pressure liberates carbon dioxide and forms caustic soda in the boiler water. The carbon dioxide, on reacting with condensate, will form carbonic acid and attack, or groove, the return lines. The high alkalinity resulting from a buildup of caustic soda (NaOH) can lead to operational problems, priming, foaming, and, under the right conditions, caustic embrittlement failure. To control this alkalinity buildup, we can pass this supply through a chloride-anion exchanger and convert 90 percent of the alkalinity to salt, or sodium chloride (NaCl). Each 264 ppm of lime will be converted to 154 ppm of salt in this process. Thus we have 233 ppm natural plus 154 ppm converted for a total of 387 ppm of sodium chloride going into the boiler per cycle of

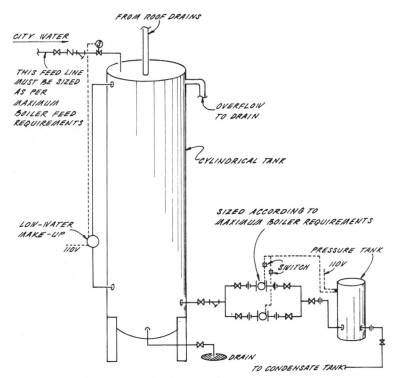

FIGURE 11   Cistern water for boiler make-up water.

concentration. Thus, if we allow 6 cycles (3500/485 = 7.2 cycles), we can accumulate 6 × 387 = 2322 ppm of salt in the water and maintain that level at a constant rate. This amount of salt translates to 1 lb per 50 gal of water, which is certainly not beneficial to the internal parts of the boiler.

The reader is at liberty to rationalize that justification for boiler water treatment is not required; however, care in selection is. Often, a specifications writer, on being informed that a water supply is hard, will specify a water softener without considering all factors and let the water treatment subcontractor take it from there. A good example is case 16300.

CASE 16300

| | Values | | |
Parameter	Raw	Soft	Cycles
pH	7.7	7.8	
Alkalinity, ppm as CaCO$_3$	215	225	3.1
Total hardness, ppm as CaCO$_3$	705	0.0	
Silica, ppm as SiO$_2$	7.0	7.0	17.9
Sodium chloride, ppm as NaCl	18.0	78.0	
TDS (by conductance), ppm	607	819	4.3

There is no doubt that softening was immediately recognized as being necessary; however, the TDS should have stood out, to the design engineer, just as the hardness did. The author is not taking into consideration the conversion of 705 ppm of CaCO$_3$ hardness to 747 ppm of soda ash (Na$_2$CO$_3$) or 282 ppm as caustic soda (NaOH) in the boiler which is punctuated by the 4.3 maximum cycles owing to the high TDS.

Let us assume that we can operate at 4.3 cycles with no operational difficulties. At this level a 150-hp boiler operating at 100 percent load with 80 percent return requires 80-gal/h of make-up water with blowdown at 25 percent of make-up water. To maintain this boiler would require many labor-hours in terms of chemical tests and timely blowdowns. In lieu of that, we can install a properly engineered TDS controlled blowdown system,* a good chemical feed system, and a twin-tank water softener, properly sized according to water hardness and anticipated water use.

A better approach would have been to demineralize, with storage during regeneration. With this approach one could operate at 25 cycles, which at 80-gal/h make-up water translates to a blowdown at 4 percent of make-up water, 3.2 vs. 18.4 gal/h, and an 82 percent reduction in blowdown with similar reductions in water treatment chemicals.

Simplistic approaches to water treatment can be quite damaging, to

---

* The author has cautioned elsewhere against believing the TDS controllers capable of reading the actual concentration of dissolved solids.

both equipment and one's professional reputation. What can one say to an owner losing boiler tubes to pitting 3 years after startup using the softened-water supply of case 54937? A good approach for a specifications writer to take, other than to become a water treatment expert, would be as follows:

1. Obtain the services of an experienced independent water treatment consultant to provide a complete, packaged recommendation for the job.

2. Provide several local, responsible water treatment firms, having experience with the water to be used at the job, with specific facts and request their recommendations. Study their recommendations and be guided by them.

3. Combine steps 1 and 2.

Of the above, the best is obviously step 3. One could obtain the services of an experienced water treatment consultant to review the recommendations of several water treatment experts. The consultant could compare recommendations and provide a rationale as to the best approach for the job.

As boiler pressures increase above 300 psig, the parameters affecting boiler operation (e.g., solids, alkalinity, silica, etc.) change. In this realm there are so many areas of concern that specialists are called in to offer specific recommendations. However, that is not to say that we cannot follow the same approach, i.e., step 3 above, to make certain we are providing the owner with the best possible program.

TWO

Closed Systems

The need for water treatment in closed systems is often not obvious — or is ignored until later, much to the chagrin of engineer and client alike. It is imperative that the design engineer fully understand *why* water treatment is needed. Then, and only then, can the engineer appreciate the problems involved and their solutions.

Water treatment of closed systems is not merely the addition of a drum of this chemical or a drum of that chemical. Instead, it is the alteration of a water supply to meet a specific need. For the professional engineer, the primary need is to insure longevity of heating, ventilating, and air conditioning (HVAC) equipment usefulness with as few repair problems as possible. The greatest enemy of HVAC equipment is corrosion, for it shortens equipment life (boiler tubes have been known to corrode through in less than a year). But the most immediate enemy of HVAC equipment is scale formation since it affects heat transfer and thus equipment efficiency — which in turn affects design parameters. It is not the author's intention to go into the chemistry of corrosion, scale formation, or germ problems; the books listed in the Bibliography at the end of this manual are excellent sources for further reading. But we will point out how equipment can be affected by such problems, and because health issues are of such importance, we will discuss bacteriological problems to some extent.

How often has one heard, "I've never used water treatment in closed systems and have never had problems. Why should I start using treatment now?" Or: "After initial corrosion takes place, the process stops and thus treatment is not necessary." Both points of view are valid to a degree — but only to a degree.

The first argument makes one wonder about the nature of the water supplies involved and the luck of the design engineer. An analysis of the corrosive tendency of the following water supply should help to explain why one supply can be corrosive while another is not. We use figures from an actual supply:

Parameters	Raw	Soft
	Source	
pH	7.7	7.8
Alkalinity, total, ppm	215	225
Hardness, total, ppm	700	14
Total solids, ppm	750	820
I_s at 40°F (4.4°C)	+0.39	−0.15
I_s at 180°F (82.2°C)	+1.77	−1.29

Calculating the saturation index,[1] we can see that the raw water exhibits noncorrosive tendencies at 40°F (4.4°C) and above since I_s is positive at 40°F (4.4°C) and becomes more so as the temperature increases.

If we were to soften the water—and with such a high hardness we certainly would—then I_s would change, since hardness is an important factor in the calculations. At 40°F (4.4°C) the soft water becomes slightly corrosive, with a negative saturation index of −0.15. As the temperature climbs, the water becomes extremely corrosive and will prove to be a significant problem to a design engineer.

If the design engineer uses the hard-water supply in a closed chilled-water system and does not experience any leaks, conceivably the system will not need any treatment for corrosion control (i.e., the system will exhibit no immediate need for attention). Some corrosion from the dissolved gases will take place after the initial fill but will, in time, be self-limiting and prove not to be too troublesome. The same basic reasoning applies if hard water is used in a closed hot-water system, with but one difference: As hard water is heated, the hardness comes out as scale on heat exchangers (HXs), and thus the water becomes soft and can become corrosive. However, as long as leaks do not develop, no problem will become evident within a contract period.

If soft water is used in the closed system, clearly corrosion will be an immediate problem, and one costly in terms of repair costs, cleanup, and ill will generated. So, under certain conditions—positive I_s, low temperatures, and a leak-free system—we could operate a system without treatment, at least for a few years or until leaks developed.

CLEANING CLOSED SYSTEMS

Once a system becomes dirty, it must be cleaned as soon as possible for the following reasons:

1. The HXs will be coated with an insulating layer of debris and thus lose efficiency.

2. The moving debris will cause erosion, the physical removal of metal, as it is moved throughout the system.

3. The debris will accelerate the destruction of pump shafts, mechanical seals, and control valves.

4. The debris will prevent any treatment from protecting the metal underneath.

5. Leaks within walls could develop, proving very expensive to correct.

The first step in the cleaning process is to determine the true nature of the dirt. Untreated closed loops will usually be loaded with black debris, rust, which may have a slight sulfuric odor. Closed loops which have been treated with nitrites or other bio-degradable organic materials may be loaded with living masses of slime. Once the true nature of the foulant is determined, one can attempt a cleaning.

The author has found the following cleaning procedure satisfactory.

Cleaning an Old Closed System Fouled with Slime

First, make certain the material is organic by taking a spoonful of it and igniting it over a propane torch. If 90 percent of the material ignites — note the odor of the smoke — and leaves a nonmagnetic residue, then you can assume the material is organic in nature.

The first step involves killing the maximum number of organisms in the system via the introduction of a suitable disinfectant. Do not confuse the terms "disinfecting" and "sterilization." The second step involves the removal of the dead material from the system.

You can accomplish the first step by injecting 3 gal of household bleach (5% sodium hypochlorite) per 1000 gal of system water to achieve a chlorine reserve of 100 ppm or more — use a chlorine test kit to ascertain the reserve — in the system for no less than 24 h. You can add the bleach via the system by-pass feeder (Fig. 12). In lieu of a by-pass feeder, you can use a hand pump (Fig. 13) or the system's circulating pump to suck in the solution (Fig. 14), if the suction side is equipped with a drain valve as shown. To suck in the mixture, connect a short hose to the drain valve and, with the pump off, open the drain valve to get the air out, while keeping the hose end below the liquid level in the bucket. Once you have done this, turn

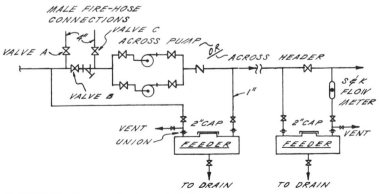

FIGURE 12 By-pass feeder installation.

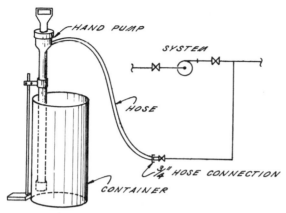

FIGURE 13 Hand pump.

the circulating pump on and throttle the valve A (the suction-side valve) until the level in the bucket starts to drop. Once the bleach has been injected into system, turn on all the circulating pumps for the recommended time, no less than 24 h; at the end of this time, you should still read a residual chlorine of 100+ ppm in the system water. The author assumes that you will make certain the system water pH stays as close to 7.4 as possible by introducing hydrochloric acid (muriatic acid) if required. During the disinfecting period, pop all drain valves to remove loosened material. At the end of this first step, drain and flush the system with city water for 15 min. Now you are ready for the second step — the attempted removal of the dead organic matter by using the alkaline cleaning agents.

Using the same method outlined for introducing bleach into the system, you now introduce soda ash or trisodium phosphate at the rate of 2 lb per 100 gal of water in the system. Instead of soda ash or trisodium phosphate,

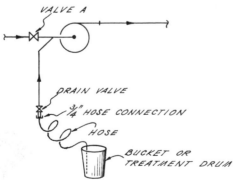

FIGURE 14 Suction injection.

Hach pH test kit. (Courtesy Hach Company.)

you could use a product specifically formulated to clean closed systems; but you must follow the manufacturer's recommendations as to dosage rates and must stick to the dosage recommended. Once the cleaning agent has been introduced, then operate all circulating pumps for at least 8 h, or as long as is recommended by the proprietary cleaning agent manufacturer, with a little heat if possible. After the allotted time, you are ready for flushing. You accomplish this by draining the system and refilling it with fresh city water; then you vent and operate the pumps for 30 min; next you drain the system and repeat the procedure 4 times, not less, until the system's water comes clean. An alternative flushing procedure would be to drain the system, with all pumps operating, at the air separator [at the Rolatrol[2] or solids separator, if so equipped (Fig. 15)] or anywhere on the discharge side of the pumps. Make certain the flush rate does not exceed the make-up water rate. During the final flush, when the circulating water no longer turns phenolphthalein[3] pink, all drain valves and dead ends should be cracked to test the flush water, which should prove neutral to phenolphthalein indicator.

After the final flush, the system is ready for treatment. Add an adequate quantity of a proper corrosion inhibitor, making certain the expansion tank also receives treatment, and maintain the level recommended by manufacturer. You can use the same method for adding treatment to the system as for adding the cleaning compound (see Figs. 12, 13, and 14).

However, when you anticipate the loss of system water (e.g., draining coils in the winter and where pump packing requires a certain bleed rate), you can utilize the method shown in Fig. 16. The constant loss of system water — 1 drop per second is equivalent to 34 gal/month — brings in a constant supply of oxygen to the expansion tank, thereby corroding the tank and supply lines. Products of corrosion can enter the system and destroy system components by their abrasive action. This problem was foreseen by two consulting firms, and the suggested approaches were utilized.

Treatments for closed systems do *not* wear out, so a constant and/or sudden drop in treatment level should be investigated and the causes of the drop corrected. Leaks of system water or bio-degradation of the product should be carefully watched and corrected before the problem gets out of hand. Remember, conditions always go from bad to worse if left uncorrected. Corrosion damage does not wait until you have the funds to fix the problem. If a bio-degradable product is used, then be prepared to sanitize the system periodically.

If the system debris fails the ignition, or burning, test and you conclude that the material is basically rust, then a cleaning procedure radically different from that just described must be used. The author has yet to hear of an alkaline cleaning compound or an organic descaling acid that will effectively clean a system of rust. Yes, these compounds do a cleaning of sorts, but you are risking radical system failure if you do not do the job right the first time. It is far better to have a system fail when you are ready for it than when the need for system operation is greatest. To clean at a cost of $600 plus your own labor, or to replace for a cost of $6000 plus — that is the question!

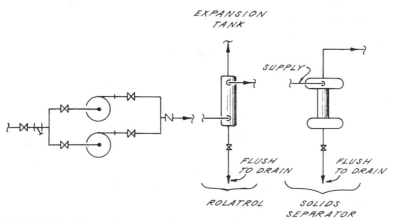

FIGURE 15 Alternative system flush.

Cleaning an Existing Closed System of Rust Deposits by Using Acid

A morbid fear of system failure and minimal knowledge of chemistry make some people cringe at the mention of acid. However, inorganic acids — muriatic acid being the one of choice — are the only acids that will clean rusted systems with any degree of success within a reasonable time (*reasonable* being defined as 2 or 3 working days).

When the term "acid" is mentioned to those contemplating cleaning a system, they immediately inquire as to both its safety and alternative methods. Let us emphasize that alternatives in cleaning chemicals are rarely, if ever, as good as the primary choice. Yes, there are questions about the use of acid, so we will address them prior to outlining the cleaning procedure.

How strong an acid solution should be used?
Use at least a 10% solution, if not one slightly higher, of a 30 percent commercial grade of muriatic acid. This translates to a 55-gal drum for each 500 gal of water in the system.

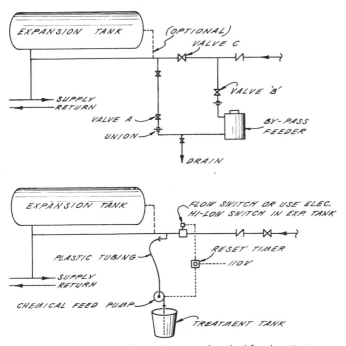

FIGURE 16 Anticipated water loss — chemical feed system.

How is the amount of water in a closed system determined?
Use the following rule of thumb. For chilled-water systems and hot-water systems with heat exchangers, multiply the pump capacity by 3; for example, $250 \times 3 = 750$ gal, so use 82 gal of acid and the rest water. For hot-water systems heated by hot-water boilers, multiply the boiler horsepower by 10. For example, for a 50-hp boiler, $50 \times 10 = 500$ gal, so use one 55-gal drum of acid and the rest water. Rather than the above guesstimate, you should use a water meter, being careful not to confuse cubic feet of water with gallons of water.

Should a straight, uninhibited commercial grade of muriatic acid, an inhibited muriatic acid, or a proprietary formulated acid be used?
Some situations require the use of uninhibited muriatic acid, but most would best be served by an inhibited grade. The proprietary formulated products's main advantage is that hopefully you acquire the expertise with the premium price paid. But if you are not going to get technical assistance and/or expert advice, why pay the premium?

What about the mechanical seals, pump impellers, or packing on the system pumps?
The fact that you are contemplating system cleaning indicates rust is already causing problems with the seals, scoring pump shafts, and reducing water flow; so the condition of these components after cleaning should not be of primary concern. Why do a mediocre job of cleaning and face system failure later? Certainly all mechanical seals should be replaced and packings renewed (three-way valves, pump packings, main valves, control valves, and pump seals on the suction side) after 2 weeks of system operation, during which time you have certainly flushed dead ends. At this point you can appreciate the value of the chamber by-pass port on the better pumps. To extend both seal and packing life, on those pumps equipped with packing, you should install cartridge filters to filter the cooling or flush water going into this chamber.

What about leaks during cleaning?
This problem should not prove to be significant. You should anticipate leaks and be prepared for temporary repairs with plenty of electrical or masking tape. Catastrophic failure is unlikely, and if the system is in such poor condition that such a failure is a distinct possibility, then cleaning it is not an option.

What about personnel safety?
When it is handled correctly, muriatic acid is safe. Certainly safety goggles should be worn, for muriatic acid is very irritating to the eyes. A fan to blow acid fumes away is also helpful, as hydrogen chloride gas is irritating to the upper respiratory tract. Since muriatic acid is not a burning acid like

sulfuric or nitric acid, however, rubber gloves and aprons are more of a hindrance than an asset. Their main value would lie in moving the drums or carboys from one place to another. If you dropped a carboy and it burst, a rubber apron would come in handy. So the object, then, is to handle the carboys carefully to make certain they do not burst.

Other important precautions include having a hose with constant running water as an emergency eye wash, a few bags of soda ash, a $2\frac{1}{2}$-gal bucket full of a 1% solution of soda ash, plenty of clean rags, tools, ventilation, and no less than two people. It is highly recommended that anyone with a history of upper respiratory problems, asthma, or emphysema not undertake the job of acid cleaning with muriatic acid. Since muriatic acid is nothing more than water with hydrogen chloride gas dissolved in it, this gas is created when a carboy is opened, which can affect those with the above-mentioned problems.

Could the system develop leaks after cleaning?
This is possible; however, with a proper water treatment program, it is very unlikely. The system should not be more prone to leaks than if it had not been cleaned.

Will the leaks be serious?
The answer depends on your definition of "serious." Certainly the collapse of a 12-in main could be considered serious — so would the rupture of a $\frac{1}{4}$-in pipe directly above a computer system! To avoid catastrophic failure, you should seek the advice of a competent consultant, who should examine a section of pipe to determine the extent of corrosion. To clean or replace — that is the question the consultant should answer. Small pipes directly above sensitive areas should be replaced.

Should problems be anticipated following cleaning?
Particulates could break away internally and cause problems with mechanical seals or continue to erode the system internally. To control or prevent such an occurrence, consider the installation of a filter, if economic conditions dictate, as shown in Fig. 17 or 18. The choice is best left to a professional engineer's judgment since one must consider the system's design parameters and be careful not to alter design considerations by reducing flow. The combination of a proper treatment program and a good filtration system should extend system life considerably.

Now that we have addressed most questions of concern, we can address the mechanics of acid cleaning. The most difficult part is getting the acid into the system, and the second problem is neutralizing the spent acid.

The easiest approach to acid-cleaning a system would be to contract the services of a competent firm that specializes in this type of work and arrives with a tank truck, personnel, and equipment to do the job. There are four advantages to this approach:

1. Maintenance personnel need not be exposed to acid solutions or fumes.

2. Experienced personnel do the job.

3. Cleanup, hydrostatic testing, and repairs can be handled by the contract firm.

4. Disposal of spent acid can and should be handled by the contract firm.

The disadvantages to outside contracting are as follows:

1. Maintenance personnel do not acquire an appreciation of the problems associated with system neglect.

2. You may have to work around the contractor's schedule.

3. There is the added cost due to unforeseen circumstances.

Should you choose outside contracting, do so with caution and deal only with a reputable firm. References are a must; or, in lieu of that, the firm should show proof of financial responsibility or post a bond in an amount equal to the risk. A written contract should be drawn up that spells out the following (plus anything else deemed necessary):

1. Definition of system, HXs, zones, expansion tanks, and condensers.

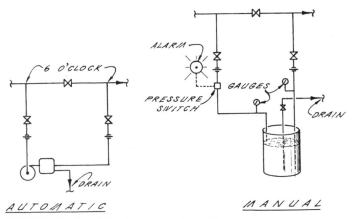

FIGURE 17 Sand filtration — less than 45 μm.

2. Who determines the final cleanliness of the system — owner, contractor, or third party? The third party could be a professional engineer, a consulting engineering firm, or a chemical consultant.

3. Method of acid cleaning and determination of effectiveness.

4. Disposal of spent acid.

5. Liability for disposal of spent acid. If the acid is dumped in a nearby lake, who pays for the consequences?

6. Final flushing and neutralizing of the system.

7. Pulling and inspecting of all strainers.

8. Replacement of mechanical seals.

9. Speed with which pipe failure is corrected.

10. Speed with which spills are neutralized and cleaned.

11. Cleanup — define this carefully! Is the contractor to take care of stained walls by painting? What about stained marble? Who does what?

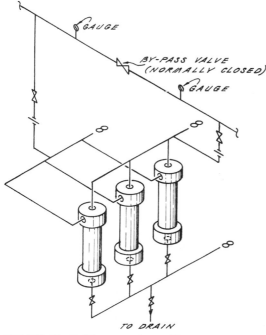

FIGURE 18 Solids separator — 45 μm.

12. Hydrostatic testing.

13. Repairs — who does what?

14. Liability for leakage causing damage to furniture, rugs, etc.

15. Who is responsible for securing acidproof tarps over sensitive equipment?

16. Who pulls the heads on HXs to check for plugged tubes?

As can be seen, acid cleaning is an involved procedure that has pitfalls. However, if the above items are discussed prior to the actual work, most, if not all, pitfalls can be avoided.

The least expensive approach is to have your own maintenance crew do the actual work. The experience gained could be applied to other areas, e.g., cleaning domestic water systems, boilers, and other recirculating water systems.

This acid-injection procedure is as follows:

1. Coat the circulating pump impeller with an acid-resistant epoxy to protect it from attack.

2. Remove flow switches, control equipment, and gauges that could possibly be damaged by acid. When in doubt, remove it!

3. Install new stainless-steel strainer inserts in pump strainers. If the system lacks strainers, now is the time to install them.

4. Install a $\frac{1}{2}$-in black iron system drain, as shown in Fig. 19, with valve and male hose end connection.

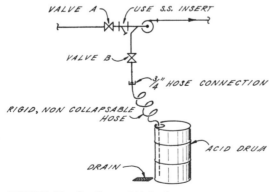

FIGURE 19 Suction acid injection.

5. Obtain a 10-ft section of rigid, noncollapsible, rubber hose to be used for suction, as shown in Fig. 19.

6. Determine the amount of water in the system — the actual metered value or the capacity, in gallons per minute, of the system pumps multiplied by 3.

7. Connect the drain hose, and drain about 15 percent of the system water. Do not forget to shut off the make-up water valve at the expansion tank or wherever it is installed. If it is at the vacuum tank, remember to vent the tank to avoid tank collapse. The preceding will require a crew of two, if not three, people to accomplish with any degree of ease.

8. With valve A open (Fig. 19) and the pump off, open valve B to fill the hose with water; then shut off valve B, making certain the hose does not empty. The hose could be placed inside the acid drum and the liquid level watched to find out whether the hose still contains air.

9. Turn the pump on and throttle the valve A until the pump starts to "complain" with an audible change in pitch, vibration, or noise. At this point insert the hose end into the acid drum, below the liquid level, and open valve B. If everything has gone according to plan, you should see the acid level drop in the drum. You could — should — practice this procedure with tap water, or even system water, to get it down pat.

The author has had firsthand experience with this method and knows it works well. If it does not work, look for the following problems:

1. The suction hose may be too long.

2. The suction hose may not be completely evacuated of air.

3. The mechanical seal may be defective, allowing air to enter.

4. The pump may be sucking air rather than fluid acid.

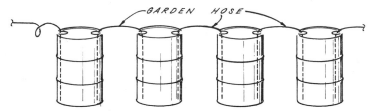

FIGURE 20 Connecting acid drums in series.

If you plan on adding more than one drum of acid to the system, it would be wise to connect the drums with syphon hoses as shown in Fig. 20. An alternative to this messy procedure would be to prepare drum taps and connect the drums with plastic tees. With the drums lying on their sides, the positive pressure assists in pushing the acid into the system. However, keep in mind that each drum weighs about 500 lb.

Is There an Easier Way of Getting the Acid into the System?
It may not be easier, but the acid pump system outlined in Fig. 21 has been used on more than one occasion with excellent results. The biggest problem encountered was the priming of the self-priming pump. The Teel 3P577 cast-iron centrifugal pump is satisfactory, both in terms of service and of cost factor. As expected, the cast-iron head does not last long — about four acid cleanings. At a cost of $60 per head, each job averaged $15 plus the cost of the nipples, hose connections, and check and gate valves. If an acidproof pump is required, consider the Corcoran close-coupled centrifugal pump model 2000-E with wetted end of Hastekloy B and C. If you are operating a service firm and can obtain acid-resistant connections (e.g., nipples, hose connections, valves, etc.), then it is certainly worth considering. For the limited once-or-twice use, the Teel unit is adequate.

Once the proper quantity of acid has been added, turn on the circulating pumps and let the system sit for 1 or 2 h. Although 2 h seems like a long time to leave acid in a system, the author has flask-tested a 20% acid solution — 20 milliliters (mL) of commercial inhibited muriatic acid and 80

Pump for acid cleaning. (Courtesy Multi-Duti Manufacturing Company.)

mL of distilled water — on low-carbon-steel coupons and has noted very little attack. The loss could not be measured with a micrometer or significantly detected on an analytical balance. However, you should experiment with pieces of system metals. The weight loss will be high initially, as the rust is removed, but will level off as bare metal is exposed.

You can judge the effectiveness of the cleaning, to a degree, by immersing a piece of system pipe section in the cleaning solution and observing the results. It is important to treat the pipe no differently from the system; e.g., water must circulate through rust, and you must not rub off the rust with your fingers. Once you have allowed the system to sit for the allotted time, you are ready to consider what to do with the spent acid solution.

If you have obtained prior approval from the local sewer operators to drain the acid, untreated, into the sewers, this is the best and easiest method of disposal. If not, you must neutralize the acid prior to dumping it.

The author has attempted neutralization in the system and has found it messy, since acid-soluble materials become insoluble as the pH is increased. Thus this method is not recommended.

It has also been suggested that one need only pour soda ash (sodium carbonate) into the sewer drain along with the spent acid. If this method is approved by the sewer operators, then by all means use it. You will require about 200 to 225 lb of soda ash for each 55-gal drum of muriatic acid used. Start off by pouring about one-quarter of the amount of soda ash required down the drain, followed by fresh water. As the spent acid is drained, add the rest of the soda ash in equal amounts. You should expect a great deal of foam and be prepared to deal with it. The author recommends the addition of a suitable antifoam, such as SAG-470, to the drain water. You should have on hand about 1 gal of antifoam for every drum of acid used. A great deal of carbon dioxide gas will be generated during neutralization, so make

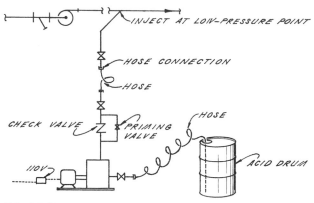

FIGURE 21 Acid pump system.

certain that

1. The vent of the sewer is not plugged.

2. Toilets are not operated during the neutralizing procedure.

3. Other drains in the plant are observed for foam and splashing of water as the gas attempts to escape.

If a storm sewer is used, make certain to get prior approval for dumping since uncontrolled acid going into a stream, river, etc., could prove undesirable from many points of view.

If a better method is desired, you could utilize that shown in Fig. 22. The soda ash is premixed and added in solution form along with the system water. You should add the antifoam directly to the soda ash solution to minimize foam.

Unfortunately, neither method provides a guarantee that effective neutralization has taken place. Their only advantage is the safety aspect, since soda ash is the neutralizing agent. The same procedure could be performed by using lye (sodium hydroxide), but this material is quite hazardous and so its use should be supervised by someone experienced with it. Its main advantage is the lack of foam.

Another method involves the set-up shown in Fig. 23. Here we use two 55-gal drums, altered as shown, and utilize lye (sodium hydroxide) as the neutralizing agent. We will require about 100 lb of lye for every 55-gal drum of cleaning acid used. The advantage of using lye, as noted above, lies in its ability to neutralize without generating foam, making it ideal for use in this application. But remember, heat will be generated, and lye can cause blindness, so the use of safety equipment is mandatory.

The following procedure should be used:

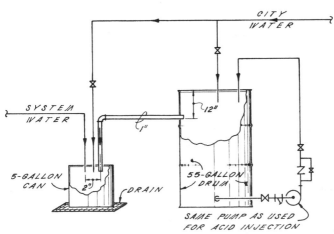

FIGURE 22   Acid-neutralizing system — soda ash.

1. Add 15 lb of flake lye or 5 gal of a 30% solution into tank 1.

2. Open the supply valve to let the spent acid solution from system into the tank until tank 2 is filled to the overflow pipe.

3. Start the neutralizing pump (the same one used to inject the acid into the system), and establish a good flow.

4. Open the supply valve and add caustic with caution, at a rate sufficient to maintain the pH of the 2-in overflow pipe within a range of 7.5 to 8.5, or slightly higher. This range can be determined by using a portable pH meter, indicator solution, or paper.

If this method seems slow, complicated, or messy, then perhaps, space permitting, you should consider using a small 10 × 3-ft swimming pool as a neutralizing tank. A pool of this size can hold about 1500 gal of water and, by using the acid-injection pump, can be neutralized quickly. After neutralization, you can use the pump supplied with the swimming pool to empty it. The question of which neutralizing agent to use in this set-up is quite important.

Lye, sodium hydroxide, caustic soda, NaOH: The heat generated with this material in the neutralizing process could prove sufficient to rupture the plastic liner. Ice could possibly be used to overcome this problem, but lots of ice would be required.

Soda ash, sodium carbonate, Na_2CO_3: This material will generate lots and lots of foam! If there is plenty of antifoam on hand — 1 percent is not an unreasonable amount — this material could be the neutralizing agent of choice owing to its safety.

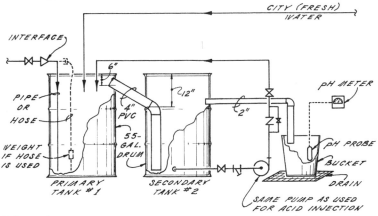

FIGURE 23 Acid-neutralizing system — lye/caustic. *Note:* Do not use this set-up for *soda ash,* because the foam may be overpowering.

Limestone, calcium carbonate, CaCO₃: This compound would react in the same manner as soda ash—you would require 60 to 70 lb for each 55-gal drum of acid used. It will not generate heat as lye will, should be less expensive than soda ash, and is easily obtained. Limestone, like soda ash, will generate a great deal of foam and, if the chunks are too large, will not react quickly. Pea size would be the best size to use. Figure on using 1 percent antifoam—about 5 gal per 1000 gal of water in the system. This may be a bit high, but it is better to have more on hand than not enough.

Once you have emptied the system, you have to flush the remaining traces of acid from it. Flush with city water—fill the system, vent and operate the main pumps for 15 to 30 min, then dump. Do the final cleaning, immediately after the flushing, with an alkaline cleaning compound, as previously described for slime-fouled systems. Once you have cleaned the system, you should treat it immediately with a suitable corrosion inhibitor. At this point, be prepared to install filters or suitable devices in order to remove traces of iron particulates; particulates are abrasive and will destroy shafts, packings, mechanical seals, and other system components via erosion. The following list may be of help:

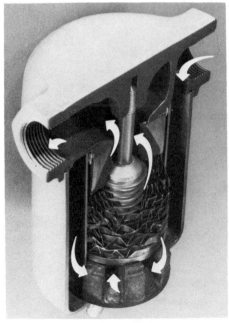

Ferrofilter PQ-3. (Courtesy S. G. Frantz Company, Inc.)

Internal view of ferrofilter PQ-3. (Courtesy S. G. Frantz Company, Inc.)

Magnetic Traps   You could consider the use of magnetic strainers, installing them on the suction side of the pump. A ferrofilter, or a dual-basket strainer, with magnetic inserts would be ideal since the baskets could be cleaned without system disruption.

Another approach would involve the installation of a series of alnico magnets in a tee (Fig. 24) properly piped. There are two drawbacks to this approach: the downtime during cleaning and the effects of a dislodged module. In lieu of a tee arrangement as in Fig. 24, you could install the magnetic modules inside the system strainers.

Centrifugal Separators   These units are being used with greater and greater frequency to keep closed and open systems clean. Although centrifugal separators are quite good for particulates larger than 45 micrometers (μm), they seem to be inadequate for smaller particulates. The author has seen a system with a bank of five units fail to clean a system loaded with particulates visible to the naked eye. Where such a condition exists, it would be advisable to add a coagulating compound to the system to make the particulates larger and thus increase the effectiveness of the separator. The author has not been too impressed with sidestream applications, so if you are considering these separators, consider a full-flow one, as shown in Fig. 18 or 25.

Filters   Because of the high pressure drop associated with filters, sand as well as cartridge, their use has been limited. However, in a just cleaned system, a sand filter in a sidestream application would be ideal. The choice of a particular unit is best left to the professional engineer who designed the system, since it is best not to alter the design parameters. As for the choice of automatic vs. manual back-flush, the author leans toward the manual units. If an automatic unit malfunctions, a great deal of treated system water could be lost before the loss became obvious.

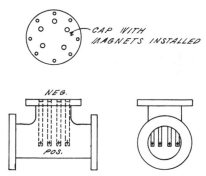

FIGURE 24   Magnetic traps.

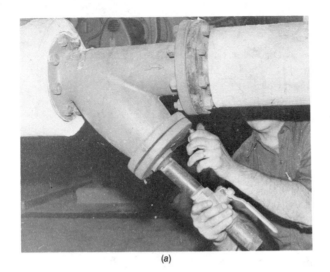

(a)

(b)

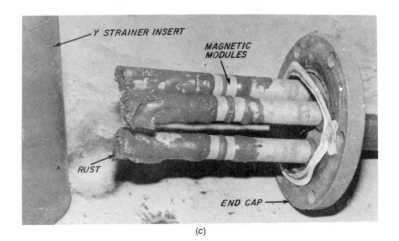

(c)

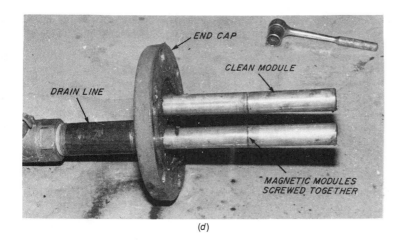

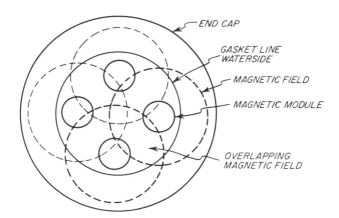

THE 1.0 X 4.0-IN MAGNETIC MODULES FROM INDIANA GENERAL EXERT INFLUENCE TO A DISTANCE 4 IN FROM MAGNET.

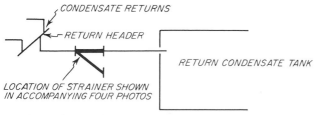

(a) Removal of Y strainer end cap; (b) end cap with rust-coated magnetic modules; (c) closeup of part b; (d) cleaned magnetic trap.

Although sand filters are superb for removing particulates down to 1 μm from the recirculating system (the human eye is capable of perceiving particulates down to about 40 μm), they require a great deal of labor power to maintain. The advertisements state that the back-wash cycle, manual or automatic, is sufficient to clean the sand of particulates; but experience has demonstrated to the author that this claim is not to be believed! Sand filters must be manually cleaned on a monthly basis by opening the unit and flushing the sand with water. As an alternative, the sand could be replaced on a monthly basis during the season. Thus it is important to specify a unit that is easily opened, via removal of the top, for sand washing or replacement.

What if the sand filter is back-washed according to directions on the unit? Within a few weeks, the differential between the operating pressure and back-wash pressure disappears. As the values of the pressures approach each other, automatic back-wash filters will back-wash with greater and greater frequency. Likewise, the manual filters will be affected, except that they will have a tendency to channel and pass all the water

Mesh screen strainer. (Courtesy Dover Corporation, Ronnin-gen-Petter Division.)

without filtration. With the automatic unit, you will waste a great deal of water; with the manual unit, you will accomplish nothing. It is evident, then, that hand washing or replacement of the sand on a monthly basis is the course of action to take.

Are there any areas where you could use a cartridge filter? Certainly! Cartridge filters can serve a vital function, the most important being the protection of the seal or packing chambers of pumps equipped with chamber by-pass ports (see Fig. 26). These ports were installed at increased manufacturing cost to provide a means of injecting cooling flush water to a critical pump area, the mechanical seal chamber. However, time and again, these ports are connected to the discharge side of the pump with a short length of $\frac{1}{4}$-in copper tubing. Although the chamber will be flushed, assuming the line is not plugged or the ports are not corroded shut, it certainly will not be with clean or cool water!

Mechanical Seals Since mechanical seal failures are due to a combination of high temperature and abrasive particulates,[4] it would be wise to filter the flush water even if you cannot cool it; however, this is not difficult to accomplish. Figure 27 shows a possible set-up to accomplish one goal, if not both.

You take system water from the pressure side of the pump (or header if

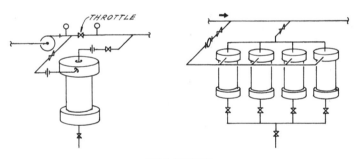

SIDE STREAM

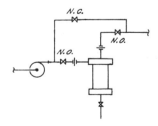

FULL FLOW

FIGURE 25 Centrifugal separators.

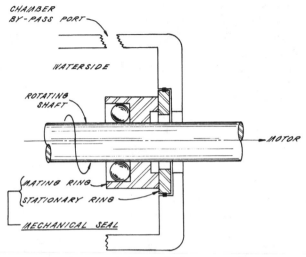

FIGURE 26 Mechanical seal with by-pass port.

more than one pump is involved) to a flowmeter or flow switch connected to an alarm—to make certain flow is present. The galvanic coupling is there to make sure galvanic corrosion does not restrict the flow. Once through the cartridge, the water should be crystal-clear. You should make certain that the cartridge can handle the system temperature, so no problem is created with disintegrating filter media.

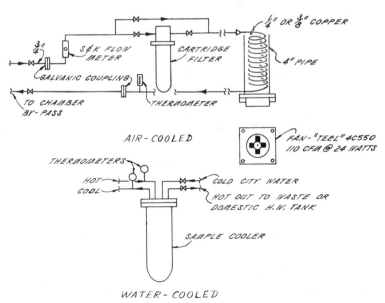

FIGURE 27 Water filtration and cooling for chamber by-pass.

From the filter you go directly to the seal chambers or through a cooling system, depending on the temperature you want to achieve. Copper tubing should not be attached directly to the injection ports, since galvanic corrosion will tend to plug the port. It is preferable to go through a galvanic coupling first and then tee off to as many seal chambers as the filter capacity will allow. There is nothing fixed about the tubing construction — it can be copper or steel — or its diameter. The only fixed item is the diameter of the pipe to the filter — it should be the same as the inlet port.

A small centrifugal-type separator is available to provide flush water to the seal chambers, but it does have a significant drawback. With most installations, the dirty water is injected back into the system, where it can erode system components, causing premature failure of zone or control valves, or allow crevice corrosion to occur where the debris builds up. Some users try to overcome this problem by dumping the dirty water down the drain, but this only compounds the problem. Every gallon of water lost from the system must be made up by fresh, cold, make-up water to avoid the bad consequences of such water loss noted at various points throughout this chapter.

NEW CLOSED RECIRCULATING WATER SYSTEM AND CLEANING IT

The time to think about a new system is at the design stage, when you should take into consideration such parameters as system temperature, source of heat (converter or direct-fired), single or dual temperature, and materials of construction. Let us begin with the specifications. Before you can add treatment, you must provide the system(s) with a proper chemical feed system and then clean the entire piping system of welding flux, dirt, and debris. The following guidelines should prove helpful.

FEEDER INSTALLATION

The following closed system___[list]___shall have installed a by-pass feeder of no less than___[2-, 5-, 10-, or 18-gallon capacity]___of suitable construction to withstand maximum system pressure. The feeder shall be installed as shown on drawing* or as recommended by the water treatment firm.

After the feeder (feeders) has (have) been installed and the system(s) is (are) ready for filling, clean the system(s) of contaminants and immediately inject a suitable corrosion inhibitor. The cleaning process and/or procedure is loaded with problems, with the question of what chemicals to use in cleaning, how and when to use them, and the adequacy of flushing

---

* See Fig. 12 for example guidelines given in this book; for guidelines used in an actual work situation, direct the reader to the correct name(s) or number(s) of the drawings.

being focal points. The author has seen specifications calling for soda ash, trisodium phosphate, alkaline compounds, and even a neutral, nonpolluting compound capable of providing a protective coating to the system metal — and the coating should last for 5 years! Apparently some people believe all the above capable of accomplishing the primary objective, i.e., cleaning the system of protective oil films applied at manufacture, welding flux, dirt, and other debris. However, whether the compound(s) will or will not accomplish the primary objective is not the point. The point is this: Why would someone with a limited knowledge of chemistry, a complex discipline, want to take on the onus of specifying a particular compound? Would it not be wiser to place the responsibility for choosing the compound on the water treatment firm and thus establish a well-defined area of responsibility? If the compound specified did not do the job, who would pay for cleaning the system a second or third time? The following guideline specification should be more than adequate to accomplish the primary objective.

CLEANING SYSTEM

The system(s) shall have installed two four- (4-) inch, male fire hose connections as shown on drawings (see Fig. X,* across header valve on suction side of pump).

The [specify] *system(s) shall be cleaned of internally applied protective oils, films, grease, welding flux, dirt, and other debris injected during construction by using a suitable cleaning agent for the task. The cleaning agent used shall not harm system components including, but not limited to, ferrous, nonferrous, plastic, natural rubber, and synthetic rubber items. Heat shall not be required for the cleaning agent. The cleaning agent shall meet federal, state, and local laws, rules, and regulations pertaining to pollution regarding disposal of flush water. It shall be the responsibility of the water treatment firm or chemical consultant to ascertain compliance. The firm supplying the cleaning agent may or may not use their own personnel for the cleaning process, but must be in active, responsible charge of the work.*

The cleaning procedure shall be as follows:

1. *The system shall be filled, vented, and leak-tested. All valves shall be open and strainers in place.*

2. *No later than forty-eight (48) hours after testing, the cleaning agent shall be added in the amount and manner prescribed by the water treatment firm or consultant.*

---

* In an actual guideline specification, give the correct name or number of the drawing; in this book, refer to Fig. 12.

3. *After addition of the cleaning agent, all system pumps shall be operated for the time prescribed by the water treatment firm or consultant.*

4. *After the cleaning period, the system shall be flushed as follows:*

   a. *City water, via valve C (see Fig. X),* shall be fed by using a fire hose. Valve B shall be closed, and valve A shall be piped to drain. In the event city pressure is insufficient for flushing, the system pumps may be used.*

   b. *After dirty water is flushed from the system, all strainers shall be pulled, cleaned, and replaced. The flushing procedure shall continue until the flush-water pH is equal to that of city or supply water.*

   c. *The final flush water shall be tested by the water treatment firm, representative, or consultant, using a test procedure capable of measuring the residual cleaning agent in the water. A pH test or time lag greater than seven (7) days will not be acceptable.[5]*

   d. *After the water has been tested and certified to be free of cleaning agent, the mechanical contractor shall submit a two- (2-) liter ($\frac{1}{2}$-gallon) sample to the owner's representative, keeping a duplicate sample for future reference.*

5. *After final flushing, all strainers shall be pulled again, cleaned, and replaced. Then the system shall be filled and treated with the approved corrosion inhibitor at the prescribed dosage. At this time, or as arranged by the mechanical contractor, the water treatment firm or consultant shall provide the owners with a set of test kits appropriate for inhibitor used, log sheets, and instructions.†*

6. *After the cleaning period, the system shall be flushed as follows:*

   a. *The system shall be drained at its lowest point.*

   b. *After the system is drained, it shall be refilled and all system pumps made operational. All dead ends shall be purged and strainers pulled at this time. All pumps shall operate for thirty (30) minutes after the last strainer is pulled, cleaned, and replaced. After the elapsed time, the system shall be drained and refilled.*

   c. *The system pumps shall be operated for thirty (30) minutes and drained again. Repeat two more times. The mechanical contractor shall test the pH of the system water prior to contacting the water treatment firm, representative, or consultant. The pH shall be equal to that of the incoming make-up water.*

   d. *The final flush water shall be tested by the water treatment firm, representative, or consultant, using a test procedure capable of measuring the*

---

* In this book, see Fig. 12.

† NOTE: Should it be impractical to install the 4-in flush connections and proceed as above, then you have no alternative but to utilize the system pumps for the flushing procedure. The procedure listed in item 6 can be utilized.

residual cleaning agent in the system water. A pH test or time lag greater than seven (7) days will not be acceptable.[5]

e. *After the water has been tested and certified to be free of cleaning agent, the mechanical contractor shall submit a two- (2-) liter sample ($\frac{1}{2}$ gallon) to the owners' representative, keeping a duplicate sample for future reference.*

7. *After final flushing, all strainers shall be pulled again, cleaned, and replaced. Then the system shall be filled and treated with the approved corrosion inhibitor at the prescribed dosage. At this time, or as arranged by the mechanical contractor, the water treatment firm, representative, or consultant shall provide the owners with a set of test kits appropriate for the inhibitor used, log sheets, and instructions.*

TREATMENT FOR CLOSED SYSTEMS

This area is loaded with variables which, if not properly addressed at the onset, can cause untold difficulties for the specifying engineer, mechanical contractor, water treatment consultant, and ultimately the client or owner. Let us take note of some important parameters that must be considered at the design stage:

Temperature

This parameter is taken for granted by everyone except the corrosion engineer. And even the corrosion engineer must call on the expertise of the bacteriologist to solve those problems associated with bacterial growth. Let us see how temperature can affect the treatment of the system.

1. With thermal expansion, the joints that did not leak at low temperatures tend to leak at elevated temperatures. Undetected and/or uncorrected leaks must be made up by untreated water, which can cause problems in the expansion tank, heat exchanger, and make-up line. Products of corrosion entering the system will cause problems with the mechanical seals.[4]

2. "Seepage," an innocuous term, is rarely given the attention it deserves. Where seepage occurs, there also occurs a concentration of mineral salts, made up of those contributed by the treatment as well as those naturally found in the make-up water supply as the water evaporates. One first notices seepage when one starts to see stalactites growing at the joints, but rarely (if ever) does one stop to consider what is occurring at the joint until it fails. What can one expect to happen at a metallic junction which is moist and laden with salt?

3. Control valves frequently fall victim to the abrasive nature of the salts as they build up on stems. The up-and-down motion of the valve exposes more mineral-laden water to the atmosphere, where evaporation causes the salts to become more concentrated and accelerates valve-stem destruction through the abrasive action of the salts.

4. Increased temperature also brings with it the increased probability of seal failure.[4]

5. Above 40°F (4.4°C) and below 120°F (48.9°C), the potential for bacterial growth increases. The following are real problems:

    a. Bacterial growth may pose a potential health hazard to the maintenance crew.

    b. Products of bacterial metabolism could accelerate corrosion and cause system failure through crevice pitting.

    c. The original water treatment product could be so altered as to be ineffective.

6. As temperatures increase above 40°F (4.4°C), chemical reaction rates also increase, and this may require an altered inhibitor program. What is effective at 40°F (4.4°C) may not be effective at 90°F (32.2°C).

System Design

System design has become so complex that it poses formidable problems to proper water treatment or system protection. The single-loop closed chilled-water, hot-water, and high-temperature systems are relatively easy to protect if the temperature remains constant. However, problems arise in the following situations:

1. Coils are drained to protect against possible freezeups in the winter.

    a. If there is moisture corrosion will occur, and the products of corrosion wreak havoc with seals, impellers, and plug strainers, as previously discussed.

    b. If glycol is injected into the coil as protection, some of, if not all, the glycol will end up in the system on startup. The glycol could react with system treatment and cause serious problems.

    c. On startup a great deal of oxygen can be expected to get into the system with the make-up water.

2. A chilled-water system converts to a hot-water system in the winter. An inhibitor package that functioned quite well at 40 to 50°F (4.4 to 10°C) may not work at all at higher temperatures.

3. Three systems — open-condenser, chilled-water, and hot-water — share a common piping system; i.e., the same water is used interchangeably depending on the desired end temperature.

4. A system is maintained close to optimum for bacterial growth at 90 to 110°F (32.2 to 43.3°C). Since the human body temperature is 98.6°F (37°C), could bacteria growing in the system pose health problems for the operating personnel? The author[6] has found bacteria growing in cooling tower water even when it is treated with chromate at 300 to 400 ppm, so it should be possible for bacteria to grow in a closed-loop system even if it is treated with chromate. If the system is treated with nitrite, bacterial growth is guaranteed. Aside from the obvious potential health problem, under adverse conditions accelerated corrosion problems are caused by bacterial products of metabolism.

5. Large heat storage tanks are buried underground. There tanks are normally associated with item 4 above, so the problems become magnified.

Chemicals for Corrosion Control

Since this is a practical approach to water treatment, we just list the popular active compounds with their pros and cons, making no effort to influence the choice of one over another. Should you desire a thorough analysis of corrosion mechanisms, study Frank N. Speller, *Corrosion, Causes and Prevention,* McGraw-Hill, New York, 1951; Herbert H. Uhlig, *Corrosion Handbook,* Wiley, New York, 1948; or *Basic Corrosion Course,* by the National Association of Corrosion Engineers (NACE), Houston, 1973 — all listed in the Bibliography at the end of the book.

Chromates   Chromates, long the mainstay of the water treatment field, are second to none for corrosion protection. The author has seen closed chilled-water systems at 45 to 50°F (7.2 to 10°C) treated at 500 to 700 ppm with no corrosion evident after 15 years of service! Hot-water systems, owing to higher temperatures, demand higher chromate concentrations, but fear of mechanical seal failure has intimidated operating engineers into using less than the required amount. However, even at 700 to 1500 ppm, excellent results have been obtained. One pump manufacturer claims even 700 ppm is too high and suggests a maximum of 300 ppm, but this is far too low to provide meaningful protection for a 180°F (82.2°C) system.

For systems above 180°F (82.2°C), the author has found dissatisfaction among operating engineers as a result of inherent system leaks. There is the never-ending debate on the yellow stain: *Did the chrome cause the leak, or did it just point out the hidden one?*

As for pollution regulations, if the system is designed to minimize leakage, uses mechanical seals instead of packing, and has pumps with flush chambers, filters for the flush water, and sufficient valves to minimize system water loss in the event of component failure, then there should be no pollution! However, some localities prohibit chromates just on principle, and others will allow them in closed-loop systems only.

Nitrites/Borates Formulas containing nitrites and borates are second to chromate formulas in extent of use in the field. For corrosion protection they are, for all practical purposes, equivalent to chrome and are nonstaining. *The main disadvantage of nitrite is that it will support bacterial growth at temperatures below 120°F (48.9°C).* Some claim bacterial growth at as high as 180°F (92.2°C); however, this is unsupported by the author's experience in such systems treated at 1000- to 1500-ppm nitrite levels.

It has been said that straight, dry nitrite/borate formulas, while providing excellent protection for the steel components of the system, offer little, if any, protection for the nonferrous components. To provide complete protection, it has been suggested that organic inhibitors be added, in the formula or as an adjunct, which are specific for the nonferrous components. Other than to consume additional treatment, the author fails to see the advantage of adding such chemicals to a system (see Chap. 7, "Chemical Safety"). How can anyone predict the metallic components of a system and thus arrive at a general formula to cover every situation? *Further, why add more organic chemicals to an already overburdened environmental system?* Having inspected and treated systems since 1963, the author has not seen — he has read about but has not seen — a single case where the use of nonferrous, organic corrosion inhibitors was warranted.

Sulfites and Hydrazine By reacting with dissolved oxygen, sulfites and hydrazine eliminate only one contributor to the corrosion process. They will not form films or stop other forms of corrosion, but they are better than nothing. Since they are recommended for systems operating above 200°F (93.3°C), where other corrosion inhibitors cannot or will not work, they are perfect where dissolved solids must be minimized or where contamination of steam boilers, through failure of an HX, is a real and present problem.

Organic Materials The list of available organic treatments and formulations is beyond the scope of this book. Suffice it to say that any properly formulated organic treatment is satisfactory as long as the following points are covered:

1. They are safe for users:

   a. Nontoxic.

   b. Not a bacterial nutrient.

2. They are environmentally safe — the products of breakdown should not pose a hazard to marine or aquatic life.

3. In the event of system failure, rupture of an HX or a pipe, they will not prove hazardous to the public.

4. The amount in the system can be tested, via a direct test kit, by the user. A pH test is not suitable for this purpose.

5. They control corrosion at an acceptable rate, and the amount of corrosion that does occur can be monitored by means of corrosion coupons.

Silicates Silicates are used very successfully in treating domestic water systems, but their use appears to be limited in closed-loop systems. The author can foresee problems with mechanical seals and valves and the use of silicates, but this is not to say that a formula incorporating this chemical should be avoided. Given the problems inherent with nitrites, one should not restrict one's options.

Molybdates Because of the high cost of this chemical group, its use has been very limited. However, with limits being placed on many of the old standbys, one can see a greater and greater use for this product. The author has had experience with the material and favors research to expand its use over that of the organic materials.

THREE

Open Systems

TREATMENT OF OPEN SYSTEMS

With these systems one is faced with three distinct, immediate problems — scale, corrosion, and the growth of living organisms — as well as problems induced by design considerations.

Scale and corrosion are controlled via the addition of suitable inhibitors, acid if needed as well as bleed-off, water-to-waste, in quantities sufficient to maintain proper parameters as close to ideal as possible. Unfortunately, this area is poorly understood by most design engineers. It is certainly asking too much to expect an engineer to be a chemist also, so most are compelled to do the following:

1. Seek the advice of a water treatment firm.

2. Use a standard set of specifications for all jobs.

3. Specify that water treatment shall be provided, but not supply specific guidelines.

As far as chemicals are concerned, there are no true secrets in the water treatment field. There may be innovative approaches to specific problems — what one pays for and hopefully gets — but the chemistry involved is well understood. One may encounter patented formulas, but a little searching should reveal other formulas exhibiting the same efficacy. In keeping with the stated objectives, the author will not be overly concerned with the chemistry, but will focus on the mechanics of maintaining proper conditions in an open recirculating water system at all times.

In seeking the advice of a water treatment firm, the design engineer could be limiting available options. More often than not, the water treatment firm will supply a set of specifications that are self-serving — limiting, if not eliminating, input from other sources. But if the design engineer has had good experience with and excellent service from the firm's representative, this approach could prove beneficial to all concerned. The design engineer could get a good set of specifications or drawings with few (if any) errors. The owner/client will be assured of having a system that works, and the water treatment firm gets an edge in bidding.

Using a standard set of specifications for all jobs is not a good practice and has proved to be disadvantageous to all concerned. To specify that an algaecide be applied to a system operating at 180°F (82.2°C) or that scale formation be controlled, etc., in a closed-loop system makes many wince at what the writer had in mind when putting together the specifications. Further, there are far too many variables — water quality, cost, discharge rules and regulations — to hope to cover even a few with standard specifications. The decision of where to place a cooling tower is a task in itself. On the roof? Parking lot? Inside? In the woods?

Stating that water treatment shall be provided and then not providing guidelines are not in the best interest of the client, design engineer, or water treatment firm(s) bidding on the work. The client may assume that an up-to-date system utilizing the latest technology is being supplied. The design engineer may assume a proper system will be installed. The mechanical contractor, the low bidder, wants everyone happy. The water treatment firm supplies a simple pump, tank, and solenoid valve for the 600-ton centrifugal machine. However, suppose that acid must be added to the water for alkalinity control and the water treatment firm does not supply it because it was not specified? Even if the water treatment firm supplies the acid, who pays for the feeding equipment and its installation? Further, who decides what is adequate in terms of feeding equipment?

Clearly, with open systems water treatment is not a simple matter. The following are guidelines, for new construction, to assist design engineers in specifying water treatment systems to meet the needs of the client. The first is a basic specification for the chemical section to be used in the program.

WATER TREATMENT SPECIFICATIONS

Intent

Water treatment chemicals for the open recirculating water system shall protect against scale formation, corrosion, algae, bacteria, and other living organisms. This shall be accomplished via the addition of ferrous and nonferrous corrosion inhibitors, scale inhibitors, alkalinity-reducing agents if analysis of make-up water shows the natural alkalinity to be in excess of 75.0 ppm, and a proper __[fill in name of -cide]__ program to control the growth of living organisms in the water. Should these organisms prove difficult to control, it shall be the responsibility of the water treatment firm to supply a sufficient quantity of a sanitizing agent, e.g., sodium hypochlorite, to sanitize the system at no additional cost to the owner. All chemicals supplied shall be provided for a period of one (1) year, twelve (12) calendar months, from date of startup, which shall be understood to imply immediately after cleaning and flushing.

All chemicals used shall be in compliance with federal, state, and local laws, rules, and regulations pertaining to discharge of said chemicals into the environment. It shall be the responsibility of the water treatment company to ascertain compliance and agree to defend and hold blameless the design engineer, owner, and mechanical contractor against any action brought against them by the aforementioned authorities for noncompliance. This section would not apply if the agents were not used in strict accordance with instructions.

The water treatment company agrees to guarantee a corrosion rate of less than five (5) mils per year (mils/yr), using procedures accepted by the National Association of Corrosion Engineers (NACE), the American Society for Testing Materials (ASTM), or corrosion coupons installed as shown in Fig. 28. If coupons are used, they shall be exposed to system water for 30, 60, and 90 days. In addition, three (3) preweighed coupons shall be supplied to the design engineer for self-evaluation via an outside concern.

Body of Specifications

Immediately after leak-testing of the system, it shall be cleaned with a chemical cleaning compound specifically formulated for this purpose. The system shall be cleaned of internally applied protective oils, films, grease, welding flux, dirt, and other debris resulting from construction. The cleaning agent used shall not harm the system components including, but not limited to, ferrous, nonferrous, plastic, natural rubber, and synthetic rubber items; nor shall heat be required for the

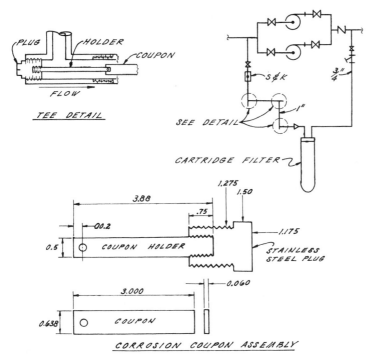

FIGURE 28   Corrosion coupon installation.

cleaning process. The firm supplying the cleaning agent may, or may not, use their own personnel for the cleaning process, but must be in active, responsible charge of the work.

Cleaning Procedure*

This procedure shall be followed:

1. *The system shall be filled, vented, and leak-tested.*

2. *Immediately after testing and repair, the cleaning agent shall be added in the amount and manner prescribed by the supplier.*

3. *After adding the prescribed amount of agent, all system pumps, with strainers installed, shall be operated for the prescribed time and flushed as follows:*

    a. *Drain the system; pull all strainers; clean, replace, and refill system.*

    b. *Operate all system pumps for no less than thirty (30) minutes, opening all dead ends, drain valves, and strainer flush valves (if so equipped) during this period.*

    c. *Drain and refill the system after the elapsed time. If system cleaning cannot be accomplished on a continuous basis during the normal work-week, it will be the responsibility of the mechanical contractor to ascertain from the supplier how best to proceed.†*

    d. *The final system water pH shall be equal, ± 0.3 pH units, to that of the make-up water. The system water shall be tested and its cleanliness certified by the water treatment company, using a test procedure capable of measuring the residual cleaning agent in the water.[5]*

    e. *After the system water has been tested and certified to be free of cleaning agent, the mechanical contractor shall submit a two- (2-) liter (½-gallon) sample of the system water to the owner's representative, keeping a duplicate sample for future reference during the period of the water treatment contract.*

    f. *After certification by the water treatment company that the system is clean and ready for treatment, the water treatment program shall be initiated. Note that the time period between final flushing and the application of water treatment shall not exceed seven (7) days.*

---

* NOTE: If the system includes individual heat pumps, sensitive equipment, computers, etc., they should be by-passed during the cleaning procedures.

† NOTE: It takes an average of four flushes, as described in subitems *b* and *c*, to clean a system adequately.

Water Treatment Program

Chemicals shall be supplied for the control of alkalinity, corrosion, dirt/debris, scale, and living organisms. Alkalinity shall be controlled via the addition of sulfuric acid, sulfamic acid, or an approved substitute, in sufficient quantities to keep the alkalinity within 350 to 400 ppm expressed as calcium carbonate. **Under no circumstances shall bleed-off, water-to-waste, be used to control alkalinity!** *Should conditions be such that the pH is forced down, by exhaust smoke, etc., then a suitable alkalinity-increasing agent shall be added, if required, to keep the pH within 7.0 to 7.5.**

Corrosion shall be controlled via the addition of a suitable corrosion inhibitor at sufficient dosage to keep the corrosion rate below 5 mils/yr. The program may be altered during the contract period to comply with specifications. Dirt/debris shall be controlled via the addition of a suitable coagulant or dispersing agent in sufficient quantities to accomplish the task. If the bidder supplies a formulated product with a dirt/debris control agent built in, this shall be acceptable. However, should dirt/debris prove to increase, as determined by visual observation, then the bidder shall be obligated to supply additional dirt/debris control agent at no cost to the owner.

Scale shall be controlled via the addition of suitable scale inhibitors at correct dosages to accomplish the task. Should hardness be the limiting factor in cycles of concentration, then bleed-off shall be utilized to maintain, in conjunction with treatment, the system scale-free.

The control of living organisms, algae, fungi, bacteria, protozoans, etc., shall be accomplished via the addition of suitable chemical agents on a schedule as set forth by the water treatment company. Oxidizing agents shall not be used on a continuous basis but may be part of the control program.†

Should scale, algae, bacteria, or fouling develop during the term of the contract, the owner having followed all instructions and maintained a proper log, then it shall be the responsibility of the water treatment contractor to supply all chemicals and labor for cleaning same. The owner's personnel may assist, to gain experience in the procedure, but are not obligated to do so. Any scale and/or growths developing in areas not constantly flooded with recirculating water shall not fall under this requirement of cleaning.

It shall be the responsibility of the water treatment contractor to ascertain the

---

* NOTE: There are claims that certain water treatment formulations work above a pH of 8.5. However, the author cautions against taking those claims seriously unless documentation is offered. A data sheet on a product is not to be accepted as documentation. If the cycles of concentration (see Chap. 1, where this term is explained) have alkalinity as the limiting factor, the addition of an alkaline treatment or a lack of alkalinity control could allow fouling and the formation of scale to occur.

† Some firms recommend the use of dip slides to check for -cide efficacy; however, the author has found these slides to be next to worthless for the intended purpose. See Ref. 6 and page 122.

fitness of the specified chemical feed system. Should alterations be required, the bidder shall so note and include it in the bid as a separate item.

Test Kits

All necessary test kits required shall be supplied by the bidder. These shall include, but not be limited to:

1. *One pH test, phenol red (pH 6.5 to 8.5), Hach No. 1470-08, or equivalent*

2. *One alkalinity test kit, Hach Model AL-AP, No. 1433-00, or equivalent*

3. *One scale/corrosion inhibitor test kit as per the water treatment contractor*

4. *One chloride test kit, Hach Model CD-50, or equivalent*

Quality Assurance

The water treatment firm bidding on this section shall have been regularly engaged in the water treatment field in the areas to be serviced for a minimum of five (5) years. The field engineer, the representative to be doing the actual work, shall have no less than five (5) years of experience in actual, direct charge of onsite testing of similar systems. Proof shall be supplied if requested. Further, the field engineer shall be a graduate chemist, A.A.S. or B.S. Testing shall be on a monthly basis, and the field engineer shall be on call, available within twelve (12) hours after being contacted to correct any emergency situations, such as, but not limited to:

1. *Chemical imbalance of system water*

2. *Chemical spills*

3. *Bacteriological questions*

4. *Occupational Safety and Health Act (OSHA) requirements and other health problems*

   Upon request, the water treatment contractor shall submit a list of at least five (5) installations of similar capacity in __[state geographical area]__ *which have been successfully treated for a period of five (5) consecutive years immediately preceding the award of this contract. Further, all submittals shall include an up-to-date OSHA product safety data sheet.*

   No system is to be treated with any chemical or compound for which there is no test procedure, the exception being the __[fill in name of the -cide program]__. *Any exceptions to specifications shall be written and submitted along with submittals. Acceptance of submittals does not waive requirements of this section.*

As can be seen, the choice of the water treatment program is left to the water treatment bidder. You, the reader, have stated what is wanted in terms of end results. It is then up to the bidder to comply, being guided by whatever laws are in effect at the time of bidding and enacted during the contract period. You have provided reasonable guidelines and have not taken the onus of responsibility on yourself by specifying a particular chemical or program. It certainly would be nice to be able to specify a particular formulated product, e.g., Venus XP-100 scale and corrosion inhibitor, for each and every job. However, just as no two separate water sources are alike, so it is with representatives of water treatment firms. It cannot be denied that the product must function, which it must do in order to survive the highly competitive water treatment market. Yet, success or failure rests with the representative's abilities to

1. Understand the water source and chemistry involved.

2. Understand the chemical program — and not burden the user with multiple additives, some of which may be of dubious value.

3. Pay attention to details and be responsive, immediately, to the needs of the design engineer, mechanical contractor, and end-user.

4. Fully understand the differences in the various heat rejectors, e.g., HXs, centrifugal units, absorbers, reciprocating units, towers, closed-circuit units, etc., and how they may affect the water treatment program.

5. Comprehend the full ramifications of the cause-and-effect relationships in chemical addition (e.g., the addition of an amine at the boiler causes an effect in the steam distribution and return-line system). This is extremely important.[7]

It is also important for the representative to know the chemical make-up of the treatment program. This, however, does pose a problem since some water treatment firms guard their secrets with greater fervor than was used to guard atomic bomb secrets. However, with the advent of the "right to know" laws, this veil has been lifted and ingredients are listed on OSHA safety data sheets (OSHA form 20). Why is this important? Let us consider one *important,* but constantly glossed over, example:

Given the following program: sulfuric acid for pH (alkalinity) control; a bio-degradable formulated product for scale, corrosion, and dirt control containing lignosulfonate, phosphonates, and other ingredients; and an algaecide program containing organic-sulfur compounds. The problem is that, even though the sulfuric acid addition has ceased, the pH of the system water constantly hovers near 2.0 to 3.0, decidedly acid. The water is rust-colored, for obvious reasons, and corrosion is evident.

In order to solve the problem, it is very important for the representative to realize that the program itself could be the primary cause. With the use of sulfuric acid the sulfate content of the water increases, helped by the addition of sulfate found in the lignosulfonate. We must consider that the organic-sulfur algaecide could be adding more sulfate via its commercial impurities and bacterial attack. Such a sulfate smorgasbord invites attack by sulfate utilizing bacteria to generate sulfuric acid in the system. Thus the pH of the system can take a nose dive even though the addition of sulfuric acid has ceased. To solve this particular problem, we must first recognize its root cause. If this is not done, all efforts will be futile. The author will not offer a specific solution to this particular problem since many factors have to be taken into consideration before it can be tackled. The reader is urged to study Chap. 4, "Algaecides/Biocides," and then review the section Cleaning Closed Systems in Chap. 2.

How about the water treatment firm's laboratory? Certainly it has qualified chemists who are capable of solving field problems. So why is it important for the representative to get involved? Well, if the representative cannot recognize, define, and explain the problem, then how is the chemist going to solve it? Further, a great deal of damage could be done to a system, especially the closed-circuit coolers, during the time that samples and letters cross in the mail.

In addition to the chemical specifications, a means of administering the additives into the system must be given. This is the area of greatest concern. The author has found situations where, for reasons understood only by the decision maker, an *adequate* chemical feed system has been installed in lieu of a *proper* chemical feed system. The term "adequate" means that the user, or specifier, *believes* it will work. Since the system worked in the next county, it will work on this job. If it does not, then, according to this decision maker, the fault rests with

1. The water treatment company, for not providing the right chemicals

2. The user, for not applying the chemicals correctly

3. The water treatment representative

4. The water supply

5. Prevailing winds

6. Proximity to roads

7. All the above

However, the term "proper" refers to a system that has been custom-designed for the specific job, with the following taken into consideration:

1. Items 3, 4, 5, and 6 above

2. Educational level and experience of personnel administering the chemicals

3. Type of system being protected, end-use

We are not saying that a specific chemical feed system is the only recourse, but rather that the system must be designed to meet the needs of the client. The literature supplied by water treatment firms touting their prewired, prepackaged units may be just the thing for one job, but not for another. Let us analyze some methods of adding treatment to open systems, in order to understand their applications, strong points, and weaknesses.

ONCE-THROUGH SYSTEMS *Forget this one*

Where you must use water as the cooling medium and where it is not suitable for reuse, you should consider a threshold treatment program. This program involves adding small quantities of a suitable scale and

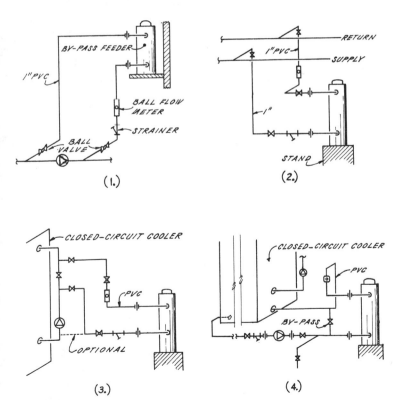

FIGURE 29 By-pass feeder use.

corrosion inhibitor on a continuous basis as long as the water is flowing. On the surface, this may appear expensive; however, most programs call for less than 10 ppm, that is, 10 lb of additive per 125,000 gal of water. The problem with this program is the actual mixing of the treatment with system water. For this reason, the actual injection point should be as far as possible from the heat-rejection source, i.e., the condenser.

The simplest approach to this program involves the use of a by-pass feeder, as shown in Fig. 29.[1] The solid treatment, e.g., a glassy polyphosphate, is added to the feeder, and the valves are set to control the water flow and thus regulate the flow of treatment out of the tank. This method would not work for a liquid product, and so is limited to solid programs.

A negative aspect of this approach is the constant attention required. As the treatment level drops in the tank, you must take one of two actions: Add additional treatment to the feeder, or increase the flow of water through the feeder.

Another drawback is the lack of real positive control on the treatment residual maintained in the system. You are forced to feed at a high rate and, hopefully, catch it before it drops below the minimum level. Figure 30 illustrates the by-pass feeder results graphically. Dotted line A is the most realistic; it will be below the minimum until someone catches it with a test on the water. It is certainly a wasteful method and quite inaccurate, but it is used because of its low initial expense.

The best approach would be to install a chemical feed pump, as shown in Fig. 31. As water starts to flow, it will trip the flow switch and activate the chemical feed pump. Note that the injection point, the corporation stop, is in an elbow (since it is an area of great turbulence) and thus assists in

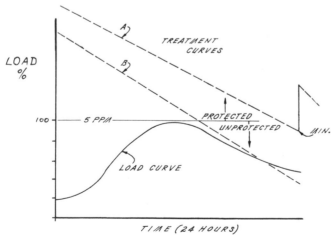

FIGURE 30 By-pass feeder use in once-through system.

mixing the additive with system water. The obvious drawback to this approach is that a constant flow through the system is assumed. Where this is the case, there is no problem; where it is not, you must regulate the injection of additive in proportion to the flow rate. You can do this by utilizing a chemical feed pump, e.g., Precision Control Code 5, capable of being paced by external means, flowmeters, or system flow regulators. As the system flow varies, so must the amount of chemical injected. The author has refrained from providing a piping diagram on this set-up because of the many variables involved (e.g., system pump capacity, flow rates, and means of controlling flow rate). A chemical-injection system for corrosion control of this nature has to be custom-designed to make certain components function with satisfactory reliability.

With once-through systems you must not forget to provide for the control of dirt/debris and organic growth. Physical material down to $45\,\mu m$ can be removed via centrifugal separators (see Fig. 18). However, do not expect these units to function well with organic matter, because the separator's efficiency is density-dependent. Figure 32 shows a recommended set-up for the control of scale, corrosion, organic growth, and dirt/debris.

The corrosion and scale inhibitor set-up, as shown in Fig. 32, is shown piped into the first elbow for aforementioned reasons. The -cide and dispersant are being controlled by means of 7-day timers which would allow no less than two on periods per week. The on period should be of sufficient duration to allow the chemical feed pumps, *sized by the supplier,* to inject a sufficient quantity of chemical into the system at a 50 percent setting. The separator also has its own timer to control the dumping cycle of the solenoid or motorized valve. As can be seen, you would not want to be dumping the separator when the -cide and dispersant pumps were in operation.

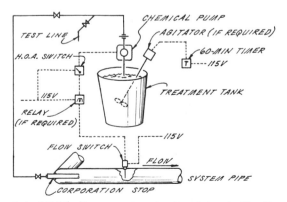

FIGURE 31 Flow switch activation of chemical feed for once-through application.

An important addition to the set-up is a 1-in test pipe, as shown in the Fig. 32 insert, across the heat-rejection source (HX) and the system pump(s). This will allow monitoring of the corrosion rate before and after chemical addition — assess the effectiveness of the treatment program. The cartridge filters here are a must since you would not want erosion (the physical removal of metal via impingement of particulates on it) to affect the corrosion rate on the coupons. The coupon holders are inserted into 1-in tees where shown. Although the insert shows two tees, consider that you may want to monitor at 10, 30, and 60 days as well as use different types of coupons (e.g., admiralty, copper, brass, etc.); thus more than the two tees shown may be required.

The turbulence chamber shown in Fig. 33 was recommended by the author to a design engineer who used it very successfully in a pH-control situation. The same set-up can be used to add treatment to a once-through system and control the pH if the water is sufficiently alkaline to warrant it. Dimensions are not supplied since pipe sizes will vary from one location to another, but a 12 × 48-in chamber should be adequate for most jobs.

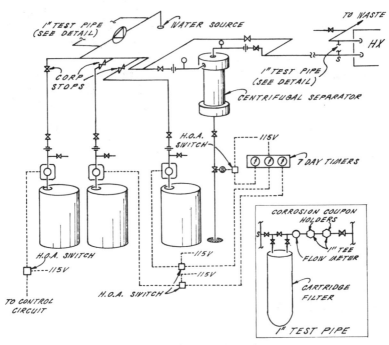

FIGURE 32 Complete protection for once-through system.

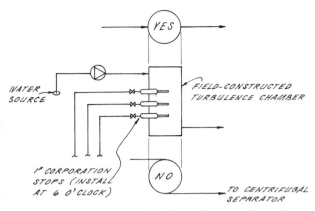

FIGURE 33 Turbulence chamber for once-through chemical treatment.

OPEN RECIRCULATING WATER SYSTEMS

The easiest approach to the addition of chemicals involves the use of by-pass feeders, as shown in Fig. 29. Simplicity and low initial cost are its strong points. This set-up may have merit with very small systems, 20 tons and under, where the make-up water rate is only 1.0 gal/min at maximum load. However, with larger systems its negative points far outweigh its positive ones, as the following analysis will prove.

Figure 34 shows the idealized situation — the load curve under the constant bleed line and below the sawtooth treatment curve. If dissolved minerals, scale, and corrosion are to be controlled, it is essential that the load curve stay below these parameters. Assuming that treatment is added

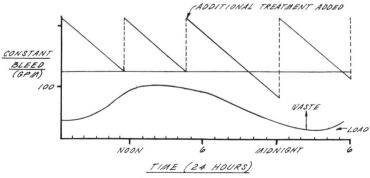

FIGURE 34 Ideal by-pass feeder treatment curve.

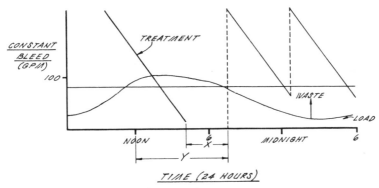

FIGURE 35 Field results of by-pass feeder use.

several times daily, based on actual onsite tests performed, you can see the following undesirable consequences of the idealized situation:

1. An excessive amount of treatment and water will be wasted.

2. The operator will be forced to spend a disproportionate amount of time making tests.

3. The environment will be subjected to excessive chemical discharge.

In actual situations the conditions depicted in Fig. 35 are found— treatment above the load curve is in excess while that under the curve is below that required for minimum protection. The time period depicted by Y represents a period of undertreatment or lack of protection. The time period depicted by X is where treatment was lacking altogether. In an effort to make things work, the operator may keep conditions as shown in Fig. 36, which is far from desirable, because items 1 and 3 above, being taken to the Nth degree, cause waste to run rampant.

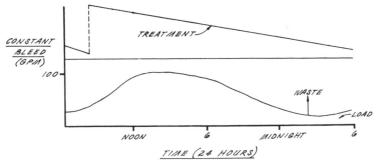

FIGURE 36 Goal of by-pass feeder use.

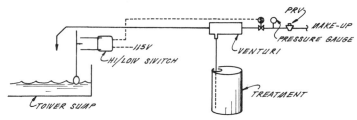

FIGURE 37 Chemical treatment via the venturi.

Besides the low initial cost, there seems to be little value in this method of treatment control. There are systems on the market that imitate this approach, and you are urged to scrutinize their presentation and/or advertisements. If there is no means of adequate control, then waste must be expected.

Simple Chemical Feed Systems

In an attempt to provide a controlled chemical feed system for towers of under 50 tons, during early 1970 the author experimented with a venturi device. At the time it was hoped that connecting the device as shown in Fig. 37 would enable controlled chemical feed. Since the venturi device works by Bernoulli's principle, a smooth, continuous flow must be maintained through the venturi. This was attempted by using a Hi/Low level switch in the tower sump to control a solenoid make-up valve. Figure 38 shows the internal parts of the venturi. As water flows in through A at X psig, a low-pressure area is created within section B. This low-pressure area is used, via the tube at C, to draw liquid into the pipe. The device functioned with some degree of adequacy, but not with consistent reliability. To make a system like this work would require further experimentation. The author is not aware of a system like this currently on the market.

A simple system that is still used in some installations, shown in Fig. 39, functions with some degree of adequacy. The 24-h timer, with multiple

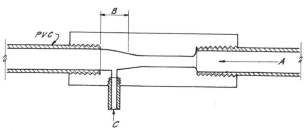

FIGURE 38 Venturi device.

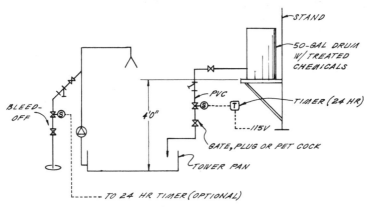

FIGURE 39 Gravity chemical feed system.

trippers, allows for a degree of flexibility in treatment addition and bleed-off, if so wired. Figure 40 depicts an operating system with a timer set to trip 10 times during a 24-h day. As the heat-rejection load picks up, shown as an air conditioning curve, the trippers activate with greater frequency, thus adding more treatment at set times. The bleed-off is shown as a straight dotted line since it shows a continuous bleed at a set rate. If the bleed-off solenoid were connected to the timer, the bleed-off would follow the same sawtooth treatment curve. Remember that the timer is a mechanical device requiring an outside source to make settings, while the heat rejector is an automatic device that rejects heat on demand automatically. Figure 41 shows two periods of demand — a hot day with a longer flat peak at 100 percent load and a cool day with a maximum load of 40 percent for a short while. If you were to place the treatment curve on Fig. 41, you would see that on that date the system would be undertreated for most of the day if it were a high-load day. With the low-load day, a waste of treatment

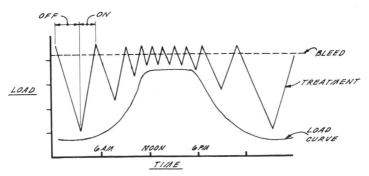

FIGURE 40 Curve of gravity chemical feed system.

would be evident. Thus, to make a system as just described function with a degree of adequacy (*adequacy* being defined as no scale formation with a corrosion rate consistent with the treatment program), you must constantly overfeed the treatment chemicals.

You should not infer from its disadvantages that the timer method is of no practical use. In some areas and situations it would be most desirable. Take, for example, an outdoor, low-tonnage (25-ton maximum) unit used for air conditioning only where bleed-off must, of necessity, be high. The following hypothetical case illustrates the point:

Water source	Analysis
Well	Hardness — 400 ppm
	Alkalinity — 100 ppm
	Silica — 10 ppm

Of necessity, the maximum concentration allowed with this supply is 2.5; that is, the water can be allowed to concentrate only 2.5 times that of the supply. At this concentration a 25-ton unit operating at 100 percent load would require a bleed-off rate of 0.5 gal/min, and at 50 percent, 0.25 gal/min. Given a 24-h/day operation at these load factors, the difference is only 360 gal of water daily, which is not very significant. As for the treatment use, by assuming a feed rate of 100 ppm, the difference is only 0.5 lb daily. Even at 100 tons you could possibly justify the simple system since water use would be at 2.0 gal/min for bleed and 3.0 gal/min for evaporation, for a total of 5.0 gal/min at 100 percent load.

Although it is inexpensive, the timer/solenoid system has been replaced with the chemical feed pump, as shown in Fig. 1, under the guise of a semiautomatic system. The timer/solenoid has preprogrammed flexibility, but the chemical feed pump, in itself, does not unless it is connected to a timer. This chemical feed system works on a continuous basis; i.e., set it and forget it. It will add chemicals to the system per the setting on the

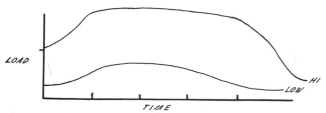

FIGURE 41 Air conditioning load curve.

pump — no more, no less. Figure 42 shows the expected curve, load, and feed relationship with this type of system. To be assured of an adequate amount of treatment in the system, it is important that you constantly overfeed as well as maintain a high bleed-off rate, above the maximum rates pertaining to a system operating at 100 percent load. Figure 43 depicts the bleed-off rates at 3, 4, 5, 6, and 7 cycles of concentration at 100 and 125 percent loads, respectively. For a 75-ton unit, other than an absorption machine, you would require a bleed-off rate (water-to-waste) anywhere within 1.1 to 1.4 gal/min, and a rate of 1.25 gal/min would be satisfactory. Once the bleed-off has been established, you adjust the chemical feed pump to maintain the treatment within recommended limits.

Although simplicity is the key to this type of system, you should understand that any treatment above the load curve is essentially *wasted*. Figure 42 reveals that an average load day wastes about 50 percent in terms of water and treatment. Is this significant? To an individual user this would not be too significant.

However, let us review an example of a 100-ton reciprocating unit used for air conditioning that operates at 24 h/day at 5 cycles. Based on Fig. 43, we would set the bleed-off rate at 0.85 gal/min and, assuming a chemical dosage rate of 1 lb per 1250 gal of water, set the chemical feed pump to feed 1 lb of treatment daily. At a treatment cost of $1 per pound, a 50 percent waste is only $0.50 per day to the user. However, let us take this a step farther and look in terms of chemicals wasted, i.e., dumped into the environment, over-all. Let us assume the average city has 100 units of this size, all operating under similar conditions and restraint. If the net waste is only $\frac{1}{2}$ lb/day per unit, we have a net loss of 50 lb of treatment daily. If the air conditioning season is 150 days, we have a loss of 7500 lb of treatment per

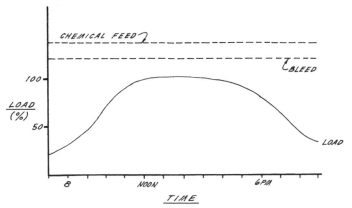

FIGURE 42 Curve of chemical feed pump and solenoid bleed-off per Fig. 1.

season. To put it another way, we are subjecting the environment, our waterways, to an unnecessary 7500-lb burden. Is this significant?

In an attempt to control this waste, we could utilize a tab-type 24-h timer to operate the chemical feed pump. Figure 44 shows just such a system with seven on periods during a 24-h day. Line *A* is the bleed-off, of necessity always higher than that required at 100 percent load to make sure there is adequate bleed when the load increases. Line *B* depicts the chemical feed on and off periods as determined by the tab setting. Conceivably, we could connect the bleed-off solenoid to the tab timer and thus obtain some measure of control over the water-to-waste. However, the desired end results would not be achieved. Figure 44 shows an ideal situation that is seldom, if ever, achieved. Remember that the heat rejector, or compressor, functions on an automatic mode and is dependent on automatic controls; thus the load curve will vary, as shown in Fig. 41, on a day-to-day basis.

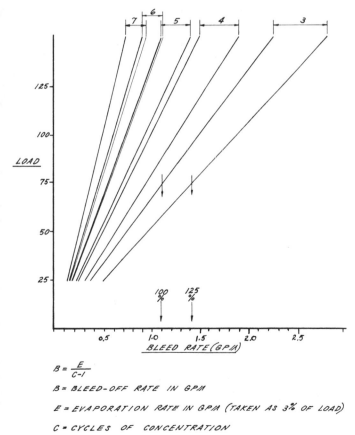

$$B = \frac{E}{C-1}$$

B = BLEED-OFF RATE IN GPM

E = EVAPORATION RATE IN GPM (TAKEN AS 3% OF LOAD)

C = CYCLES OF CONCENTRATION

FIGURE 43   Bleed rate vs. cycles of concentration.

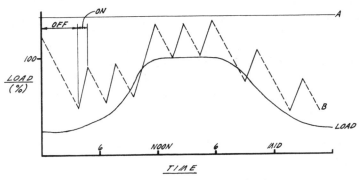

FIGURE 44 Tab-type timer-activated chemical feed pump.

Since the tabs are mechanical devices, they will always control at the level set; thus there will be periods of inadequate treatment and bleed, which will allow insulating lime to accumulate on the heat-rejecting surfaces. If the bleed is adequate and the treatment inadequate, then we could be experiencing higher than acceptable limits of corrosion. In an effort to tie the bleed-off and chemical feed together, additional automating equipment has been added to the basic simple system to correlate these parameters to the actual load.

Automatic Systems

The first step in automating, if you do not accept the simple system shown in Fig. 1 as automatic, is the use of dissolved-solids controlling devices, TDS units.

Theory holds that as the heat rejection on a system increases, the evaporation at the heat rejector, the tower, also increases in proportion to the load. As evaporation climbs, so does the concentration of dissolved minerals in the water since solids do not evaporate at cooling tower tempera-

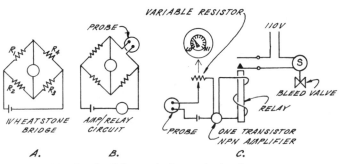

FIGURE 45 Conductivity controller — electronics.

tures. Given this consideration, an attempt is made to control the total dissolved solids and the treatment via conductivity controllers.

Many TDS units are in use, and it is unfortunate that often the user does not understand their operating principles. With the advent of the space age, computers, and integrated circuits, an electronic mystique surrounds these devices, and the user is led to believe the TDS unit capable of more than just reading the electric resistance of the water. You should understand that these units, no matter how they are packaged, are just basically ohm meters, perhaps Wheatstone bridge circuits (see Fig. 45). Part *A* shows the basic circuit; when R_1, R_2, and R_3 are known, you can calculate R_4, via the meter M, using the formula

$$R_4 = \frac{R_1 R_3}{R_2}$$

Part *B* shows how the circuit could be used to measure, or control, the resistance, or conductance, of the water. The cell, or probe, is exposed to the system water; and via other parts of the circuit, i.e., the amplifier and relay, the probe is used to control or measure the conductance, or TDS, of the water. The controller may be calibrated to read in units of TDS or of conductance, i.e., the micromho.* The term *micromho* is defined thus:

$$\text{Micromho} = \frac{1,000,000}{\text{resistance, in ohms } (\Omega)}$$

That is, where resistance is equal to 100,000 Ω,

$$\text{Micromho} = \text{conductance} = \frac{1,000,000}{100,000} = 10$$

Part *C* is a nonmeasuring, simplified controlling circuit that does not use the Wheatstone bridge circuit. The variable resistor, a potentiometer, is set to the desired conductance. When that conductance is reached, the circuit is completed; then the amplifier, an NPN transistor circuit, pulls in the relay, thus activating the control circuit, of which the bleed valve is a component.

As can be seen, the circuits are simple and have been tested by time; they are not out of the space age, futuristic, or unfathomable. The probe cannot "see" or measure anything other than the conductance of the water, and then only through its circuitry. The heart of the entire TDS controller system rests with the probe. Early units had the electrode surfaces in the

---

* In some fields, the *sieman* (S) is the preferred terminology for the mho; but "microsiemens" is so rare in the HVAC field that we have retained the more widely used term "micromhos" instead.

probes plated with platinum to make certain the cell constant did not change because of electrolysis, corrosion, or abrasion. The probes on today's units are made of chrome-plated steel or graphite. Some are easy to clean, and others are impossible! Since it is vital, the probe must be kept clean!

The TDS controlled automatic system is depicted graphically in Fig. 46. As the load (dotted line) increases, the solids (the broken line on the TDS curve) increase also. When the conductance reaches a preset high limit (the TDS on range), the controller activates a solenoid valve to dump water from the system at a prefixed rate, A, B, or C, until the conductance is lowered to the TDS off range. The bleed-off rate must be set so that when the solenoid is activated, the conductance will be lowered within a reasonable time limit and will not remain in the on mode for an extended time. Since the treatment reserve will be affected by the TDS controller, it is important that the chemical feed pump be set high enough to compensate for the bleed-off. Clearly, the graph is a simplification of what actually goes on. In practice, the on/off spread will be closer than that depicted, perhaps on 10 min and off 10 min at 100 percent load spread.

Now that you have some idea of what a TDS controller is and what is expected of it, let us analyze the over-all picture and determine the deficiencies of the program.

Since the TDS controller is basically an ohm meter, the only compounds it can read or measure are those which have the ability to pass current in water. For example, for salt in water, the higher the concentration, the greater the current flow, or the lower the resistance. Pure distilled water will not pass a significant amount of current and thus will show a very high resistance. Also, the nature of the compound must be considered. Organic compounds (e.g., aspirin, alcohol, and glycol), since they do not ionize, for all practical purposes cannot be detected, i.e., read, by the TDS controller.

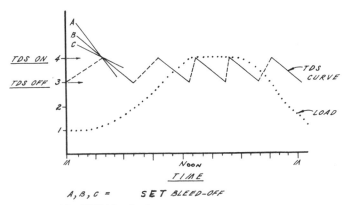

FIGURE 46   TDS bleed rate curve.

If you have access to an ohm meter, or TDS controller, the following experiment should prove to you that such an instrument cannot detect organic compounds:

1. Add 1 oz of salt to 5 gal of water.   Reading = _____
2. Read tap water.   Reading = _____
3. (1) minus (2) is the amount added.   Reading = _____
4. Repeat steps 1, 2, and 3 with aspirin, sugar, etc.

The author, using a Hydac conductivity tester from Cambridge Scientific Industries with automatic temperature compensation, ran the experiment, using city tap water, table salt, and an aspirin tablet. The aspirin weighed 355.4 milligrams (mg), and an equal amount of salt, 355.4 mg, was weighed out. The amounts, when added separately to 1 liter (L) of water, made up solutions containing 355.4 ppm of each substance. The following readings were obtained:

	Aspirin	*Salt*
	300 micromhos	1000 micromhos
Less city reading	−325 micromhos	−325 micromhos
Net reading	−25 micromhos	675 micromhos

Analysis showed that the city water alone had a resistance of 3077 Ω. On adding a salt to the water, the resistance of the water should have decreased, or electricity should have passed more easily; and that is so since $1,000,000/1000 = 1000$ Ω. When salt was added, the resistance of the water decreased from 3077 to 100 Ω! However, similar results were not obtained with the aspirin, for there was no significant change in conductance.

Another point to consider is the inability of the TDS controller to differentiate between solids. If you make up a solution containing 1000 ppm of sodium chloride, 1000 ppm of sodium carbonate, and 1000 ppm of trisodium phosphate, you will not obtain a reading of 3000-ppm total dissolved solids. Although there is a direct relationship between conductance, in ohms or micromhos, and concentration of a pure salt, this relationship does not hold for a mixture. This is an important point, and one most difficult to grasp. This can be likened to your being asked how many cars fit in the colliseum without being told how many were small, compact cars and how many were airport limousines. If the size of the exhibition space is known, you can calculate how many compact cars there could be, but not a mixture of both.

Based on what has been covered, clearly the TDS controllers have certain limitations. The greatest is that they cannot read or measure total dissolved solids of an open recirculating water system. This is most important! Since cycles of concentration are based on a precise knowledge of the mineral make-up of the water — alkalinity, silica, and hardness — you should use a precise method to determine the cycles of concentration in the water and not rely on the TDS controller. The controller will sense an increase in solids as system water evaporates; it will also sense the inorganic solids contributed by the treatment used as well as those gleaned from the air by the washing action of the tower. It will not, and cannot, detect algae, bacteria, or organic contamination such as from grease, oil, or kitchen exhausts.

We have already said that the heart of the TDS controller, based on circuitry, is the probe. It has a precise nature. The diameter of the exposed electrodes and the distance between them are built-in constants which must not be disturbed. However, with an open recirculating water system, we have built-in disturbances — dirt, slime, and all manner of encrusting debris. With clear, clean water the probe should function for a few weeks without attention; however, this is seldom the case. Cooling tower water is such that it will foul, or coat, the electrodes and thus change the constant. This coating will cause the circuitry to read a higher resistance (purer water) and thus limit bleed-off. When that happens, we will have an unexplained layer of scale on the tubes even though "we kept everything right on the nose."

Do the disadvantages outlined imply that TDS controllers are worthless? Not at all! The author simply wants to point out their inherent limitations so that you are not deceived into believing the units capable of doing more than they can. A Wheatstone bridge is a Wheatstone bridge, and it is capable of measuring resistance only when certain conditions are met.

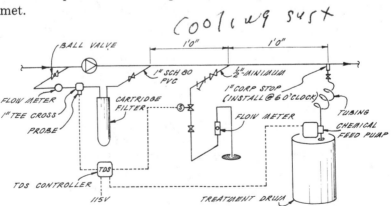

FIGURE 47 Chemical feed via conductance controller.

To use or not to use a TDS controller for controlling the important parameters of scale and corrosion in an open recirculating water system is a choice left to the reader. Should you decide that such a system is viable, then refer to Fig. 47. The probe has been placed in a by-pass arrangement around the circulating pump in a 1-in cross tee. Ahead of the probe is a cartridge filter; a strainer will not suffice. Downstream of the probe are a flowmeter and a regulating ball valve. The following is a sample specification:

CONDUCTANCE-CONTROLLED CHEMICAL FEED/BLEED SPECIFICATION

The chemical feed system shall be piped as shown in the drawings, with no exceptions, and shall consist of the following.

1. *The cartridge filter shall be a rust and dirt type of cold-water unit. It shall be of clear glass construction, or clear plastic, and capable of withstanding 125 psig at 100°F (37.8°C). The design unit is a Teel 1P635 with a 1P753 cartridge; it measures $12\frac{1}{2} \times 4\frac{1}{2}$ in with a $\frac{3}{4}$-in normal pipe thread (NPT) inlet and outlet. A sufficient quantity of replacement 1P753 cartridges shall be provided to enable the operating engineer to replace the cartridge every two (2) weeks.*

2. *The TDS controller shall be a weatherproof model and designed for cooling tower use. The electrode material shall be graphite and easily rejuvenated without the need for wire brushes or special tools. The electrode shall be capable of being inserted into a standard one- (1-) inch cross tee as shown in drawings. The controller shall not require greater than 110 volts (V) ac, and the output current shall be no less than five (5) amperes (A) in a fused circuit. The design unit is a Morr ACU/TROL conductivity monitor, controller model JA-1.*

3. *The flowmeters shall be of same manufacture and shall be as follows: For a TDS circuit, the flow water shall be no less than one-half ($\frac{1}{2}$) inch and have a flow rate within one to five (1 to 5) gal/min. For a bleed-off circuit, the flowmeter shall be capable of a flow rate sufficient to maintain proper parameters as determined by the water treatment contractor. The design unit is an SK ball flow indicator.*

4. *The chemical feed pump shall be properly sized for feeding out of drum. If this is not possible, then a fifty- (50-) gallon plastic tank with a fiberglass cover shall be supplied. The chemical feed pump shall have the motor and gears sealed in oil with no exposed moving parts. The design unit is a precision positive-displacement 8000 series pump. However, an mRoy or a Neptune DiaPump will be accepted as equivalent.*

Figure 47 is quite straightforward. The by-pass around the pump should be Schedule 80 polyvinyl chloride (PVC) or chlorinated polyvinyl chloride

(CPVC) if PVC is not available. You should remember to specify galvanic couplings when black iron is used, since the valves and flowmeters are made of bronze. This line is specified, in the drawing, as 1 in to make certain of the flow. You will have to use reducers at the filter and flow-meter. The bleed-off line should be no less than $\frac{1}{2}$ in, and $\frac{3}{4}$ in would be better. The distance between the by-pass, bleed, and injection points should be 1 ft to prevent one from interfering with another. The by-pass and bleed-off nipples should be at 9 o'clock whereas the injection point should be at 6 o'clock. The chemical feed line can be the tubing supplied with the pump, or Schedule 80 PVC may be used. If PVC or CPVC is used, you are urged to read the section on acid feed (see discussion of Fig. 56, later in this chapter) and use the same installation procedures here.

This TDS controlled chemical feed system works as follows. The conductance of filtered system water is sensed by the probe/controller as the water passes through the pipe. The flowmeter is used to set the flow low enough to extend the life of the cartridge and to make certain flow is present. The TDS controller is set below the maximum setting—you would do well to black out any reference to TDS or conductance on the dial and write in numbers from 1 to 10, at which the controller comes on. On a daily basis, reset the controller based on the results of chemical tests performed on the system water, e.g., cycles of concentration by using the chloride test. At some point in the setting, the probe reading should correspond to proper cycles of concentration, as performed by tests. Lower the setting slightly below that point, and monitor the progress, or the control function, via chemical tests. Since the chemical feed pump is connected to the controller, the controller will tend to try to "catch its own tail," which explains why the daily chemical tests are a must. The TDS controller cannot differentiate between solids contributed by normal evaporation and those gleaned from the air or contributed by the treatment.

Note how the TDS controller can fail if it is not properly maintained:

Failure during on mode: In this mode, the bleed-off stays on, the solenoid valve remains activated, and the chemical feed pump continues to pump until the tank runs dry. The consequences are obvious: (1) The system

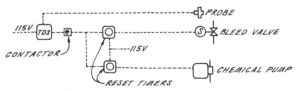

FIGURE 48 Conductance controller with two reset timers.

gets loaded with treatment, which could prove costly depending on nature of treatment; and (2) within a short time all the treatment will be bled from system and into the environment, drained. If this goes unnoticed for a long time, the recirculating system will be unprotected and scale/corrosion could take its toll. If you are feeding treatment from the drum and have no spare, you will be forced to operate the system without treatment until the next shipment. At $400 to $500 per drum, this type of failure could prove costly.

Failure during off mode: In this mode, there will be no bleed and no feed of treatment. Scale will certainly build up, and heat transfer will drop.

To prevent failure, it is essential to maintain a program of daily testing. The probe should be kept very clean and maintained per manufacturer's instructions. The handles on the by-pass valves can be removed and hidden to prevent accidental closing. Also, do not forget to pull the strainer on the bleed line and clean it at least monthly.

In an effort to improve reliability, you could alter the basic TDS controlled chemical feed/bleed system as shown in Figs. 48 and 49.

In Fig. 48 the TDS controller is used only as an initiating device. The TDS controller is used to pull in the contactor, which functions only as an on/off switch, and activate the two reset timers. The timers will operate the two parameters, chemical feed and bleed, independent of each other, thereby allowing precise control. This set-up appears satisfactory, but it is subject to the same problems already mentioned should the TDS controller fail: If it fails during the on mode, the reset timers will not reset; if it fails during the off mode, the reset timers will not come on. Either way, you will not be adding chemicals or bleeding the system. If this problem remains uncorrected, scale will build up!

In Fig. 49 the TDS controller is used only for controlling the bleed-off valve. The treatment is being controlled via a 24-h tab-type timer. This set-up guarantees that the chemical feed will proceed independent of the TDS controller. Is this any better than the above arrangement? It is differ-

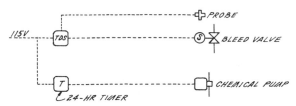

FIGURE 49 Timer-activated chemical feed system with conductance bleed.

ent, but not better. If the TDS controller fails during the on mode, the bleed rate will be very high and so water waste will be high. Since the chemical feed pump remains on, the system is somewhat protected. Should the controller fail during the off mode, the chemical feed pump will continue to operate and add chemicals to the system. If this goes uncorrected, the possibility of corrosion (if the treatment is on the acid side) or scale buildup (if the treatment is on the alkaline side and not capable of scale control under these extreme conditions) is very real.

In all the failure modes, the author does not take into consideration equipment failure, only failure of the probe, the heart of the controller. The probe can fail from the cord being cut, from the coating on the electrode, or from inadvertent shutoff of the water flow. In anticipation of this problem, you could specify a TDS controller with an alarm that would sound in the event the high TDS condition was not corrected within a set time. The Morr ACU/TROL specified earlier is not such a unit. Further, to guard against water flow failure, you should install a flow switch connected to an alarm. The author is partial to the McDonnell Miller FS4-3 flow switch; however, any similar flow switch should prove satisfactory.

Obviously the cost increases every time you attempt to upgrade the TDS chemical feed/bleed system. The basic system requires a flowmeter and cartridge filter in a properly installed by-pass arrangement. This is the minimum. Operating engineers may be told that this is not necessary with the particular piece of equipment being offered, but what has been discussed is reality. It is up to you to assess the merits of the TDS controller being offered based on what has been gleaned from the information supplied here.

Is there an alternative to the TDS controller that will accomplish the same task at reduced cost and with improved reliability? The author believes there is — in the form of a system utilizing the make-up water as a means of control. The make-up water can be monitored via a Hi/Low level switch in the tower sump or via an electric-contact head water meter. In either case, chemical and bleed control can be achieved based on a known volume of water.

Figure 50 depicts a system using make-up water for activation. Part *A* shows a short time span, 1 h, during a period of maximum load. Note that during this period there are six on activations of the timer, which has been set to activate every 6 min and then reset. The solids increase, on the other hand, varies with load and is shown climbing from 6 to 2 min at 10:52 a.m. In part *B* there is an expanded time frame, 6 a.m. to 12 noon. Here the activation periods vary from two to five per hour as the load, or evaporation at the tower, increases. Notice that the feed/bleed sawtooth curve is always above the load curve; this is designed to make certain the circulating water system is always protected by maintaining an adequate treatment level and keeping a proper bleed just slightly above the load curve (remember, anything above the load curve is waste). However, the area under the load curve represents poor conditions with respect to treatment and bleed. The operator controls how close one gets to the load curve.

The solid line covering the sawtooth curve in *B* represents parameters that the chemical feed and bleed-off system cannot exceed because of the mechanical nature of the feed/bleed system. For every known volume of water, *X* quantity, a reset timer is activated for *Y* time. The upper limit, high solids, is controlled by the water level, or volume added to the sump; and the control point, or lower level, is controlled by the setting on the reset timer. Although Fig. 50 shows the activation time at 6 min, it can be set to

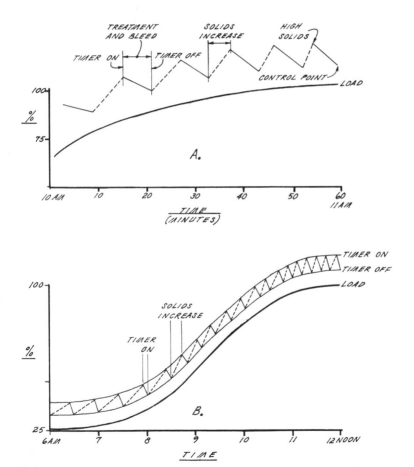

FIGURE 50 Graphs of bleed-off and feed using make-up water for activation.

any time, as determined by the water treatment contractor based on parameters important to the treatment program. However, at no time should the activation time be set so that the activator trips on itself; i.e., the reset timer must not be activated until it has reset, because the timer will not acknowledge the activation.

There are too many variations of the bleed/feed automatic system, using make-up water for activation, on the market for the author to cover all of them fully. Thus only two types will be presented here in depth, and two others discussed cursorily.

Direct Float or Level Switch Activation This method, shown in Fig. 51, uses a float or level switch in the tower sump to activate the reset timer. Once activated, the timer pulls in the relay, through the hand-off-auto (HOA) switch, thus activating the make-up solenoid, bleed-off solenoid,

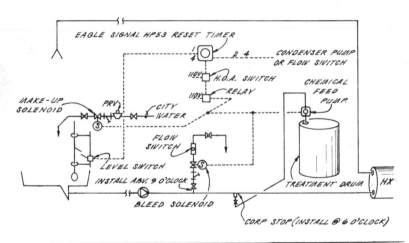

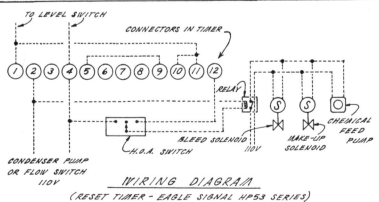

FIGURE 51 Level switch activation for bleed-off and chemical feed.

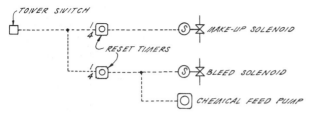

FIGURE 52 Level switch with two reset timers — one for make-up solenoid and one for bleed solenoid and chemical feed pump.

and chemical feed pump. The sump level can be controlled by throttling the ball valve after the make-up solenoid. The pressure-reducing valve (PRV) should help by maintaining a constant pressure in the feed line. We could improve this system by adding an additional reset timer, as in Figs. 52 and 53. Both these are feasible and will function with reliability. We could even go to the extreme of adding a third reset timer and have each control point independently controlled.

Indirect Float or Level Switch Activation This method utilizes an electronic water meter with a set volume pulse. The float switch, a McDonnell Miller series 80, trips an adjustable timer [Agastat 7012AE, 20 to 200 seconds (s)] to open the make-up solenoid for X time. After a calculated volume of water has passed, the water meter will send a timed pulse to the chemical feed pump and bleed-off solenoid. During this timed pulse, an adequate amount of treatment is added to the system, and a sufficient amount of water is drained. Figure 54 shows this system set-up for an indoor sump.

Direct Water Meter Activation Here we have a water meter as the primary initiator of the treatment program. After a precalculated amount of water has passed through the water meter, an electric switch inside the

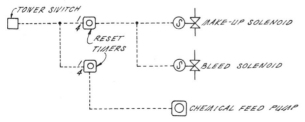

FIGURE 53 Level switch with two reset timers — one for the two solenoids and one for the chemical feed pump.

water meter closes and completes the circuit across terminals 1 and 4 of the reset timer. The timer, now activated, pulls in the relay through the HOA switch for the set time interval. The relay then activates the bleed solenoid, treatment, and acid or alkali pumps. After the set time, the reset timer resets itself and is available for activation once again. Figure 55 shows this set-up as well as the wiring diagram for any size recirculating water system.

In comparing the water meter system with others, we naturally want to know "what if." Let us review each point of potential failure and ascertain the consequences.

Water meter Water meters are fairly reliable and operate for years without failure. The author has used several makes over the years and has found the Carlon units to be very good. The gears, made of plastic and enclosed, require no lubrication or attention. And, because of their slow rotation, the gears last a long time. However, even the best can fail. If this should occur, then the system becomes inoperative — no bleed or feed. Even if the water meter should fail while tripping the timer, the system would still fail in the off mode. Since you make daily tests, you would catch this.

Reset timer The author cannot describe the failure rates of reset timers other than Eagle Signal units — and even then has never personally known of an Eagle Signal unit to fail. The only way this unit can fail is either to fail

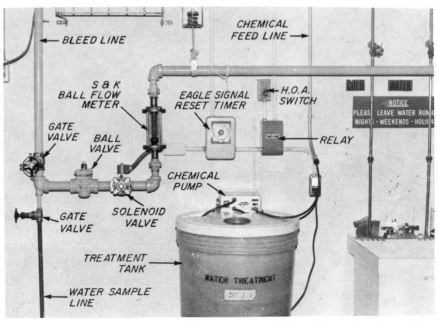

Chemical feed and bleed-off.

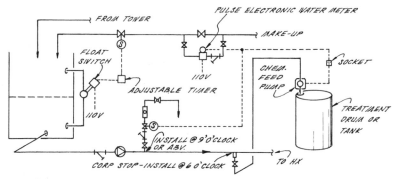

FIGURE 54 Chemical feed via pulse water meter.

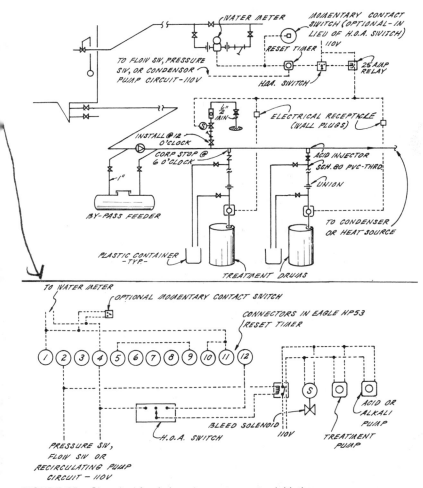

FIGURE 55 Chemical feed via pulse water meter initiation.

95

to reset or to cease functioning altogether. Should either event occur, the system will not feed chemicals or bleed. The author knows of no way the timer can or will fail in an on mode. Failure happens in the off mode.

Hand-off-auto switch   This is the problem area! The switch itself will not fail, but it could be left on in the hand, or manual, mode. This will cause chemicals to pump continuously and the bleed to stay open. With this failure, you can dump a whole drum of treatment chemicals into the system in a short time. Could this be a good argument for not feeding out of the drum? Where sulfuric acid is used for pH control, as shown in Fig. 55, you would be wise to discard the HOA switch in favor of the momentary-contact switch, as shown wired across terminals 1 and 4 of the reset timer.

25-A relay   This relay is used to protect the contacts in the reset timer, rated at 10 A, from premature wear. It is also good practice to include this relay since you never know the amperage rating of the chemical feed pumps. For example, Neptune DiaPumps draw in excess of 5 A, as do other large chemical feed pumps; so putting two of them together, one for acid and one for treatment, could burn the contacts in short order. It is easier and less expensive to replace the relay than the reset timer.

What about the direct and indirect float switch methods? In Figs. 51 and 53 the author can see no real problem resulting from failure that is not common to other systems. All three systems, if properly maintained, will function with equal efficacy.

You could improve the systems shown in Figs. 54 and 55 by providing separate reset timers to control bleed, treatment, and acid feed. However, this would only complicate matters. With the basic water meter system (Fig. 55), you have to keep in mind the following:

1. Making the reset timer higher or lower will affect the bleed-off and treatment feed similarly. The timer should be the mode of last resort.

2. Increasing the amount of bleed-off at the bleed-off valve will require an adjustment at the chemical feed pump.

3. Decreasing, or increasing, the treatment feed normally will not necessitate adjustments elsewhere.

Midcourse corrections to the water meter system are indications of problems that should be looked into and solved. The water meter system, if properly designed, is so perfect that any variations are a reason to look for the problem. For example, if the TDS value climbs, do the following:

1. Make certain no one has cut back the reset timer to reduce treatment.

2. Check for proper amount of bleed-off (in gallons per minute); then check and clean the *Y* strainer ahead of the bleed solenoid.

3. Ascertain the nature of the solids — are they treatment or make-up water salts?

4. Check whether someone has opened the by-pass valve on the water meter.

Thus additional timers, though very nice on paper, can complicate matters for the operating personnel. The easier something is to do, the greater the likelihood that it will get done, so why make things more difficult? Other improvements to the basic water meter system include using TDS controllers as primary and secondary controlling devices. The water meter controls the chemical feed pump, and as a primary device the TDS controller controls the bleed-off valve. As a secondary device it functions as a fail-safe device, coming on to activate the bleed-off solenoid valve, overriding the reset timer in the event that the water conductance exceeds a preset level. To understand the author's lack of appreciation for such devices used in conjunction with a water meter system, remember that TDS controllers only measure, or control, conductance, which is not indicative of lime-forming conditions. Admittedly, high conductance, a vague term at best, could indicate corrosive conditions (see the discussion of Langelier's saturation index on pages 2–5 for the contribution of high solids to the corrosive index), but such conditions are usually caught by the daily testing program and proper upkeep. Nothing can be so automatic that the human element is eliminated altogether.

The most important improvement to the water-meter-activated chemical feed system, if water analysis requires it, is the addition of a pH- or alkalinity-controlling pump and tank. Figure 56 shows a recommended installation along with proper safety precautions. The acid injector must be installed at 6 o'clock since acid-resistant check valves are not normally equipped with spring-loaded check valves. Immediately after the check valve, install a vent pipe with a PVC valve for venting. The vent pipe should be connected to a drain or be capped to prevent accidental spillage. This vent is used to assist in draining the acid-injection piping in the event that the acid pump requires service. To service the acid pump, shut off the valve connected to the acid injector at the pipe. Then open the drain valve, and catch the acid in a pail or let it drain back into the drum or acid carboy. Caution should be used at this stage — do *not* consider draining the acid back into the carboy or drum if concentrated sulfuric acid is used! Once all draining has ceased, shut off the valve immediately above the union and disconnect the pump at the union. Then you can take the pump to a sink for flushing and draining and to a table for service. Pay particular attention to the shims noted under the pump — this is most important! The shims are used to brace the pump as the union coupling is tightened. If shims are not used, there is the very real possibility of ripping the discharge valve from the head of the acid pump. This would necessitate the replacement of the pump head (costing $50 to $150) and the discharge valve (costing $25 to $75), since reuse of the damaged parts could be dangerous.

Is there an alternative to the use of Schedule 80 PVC threaded pipe? If you are considering low-pressure areas of 50 psig or so, you could use high-density polyethylene tubing, the same tubing as used for the suction line, as long as the run is kept short, for concentrated sulfuric acid is quite heavy — 16 lb/gal. If dilute sulfuric acid is used, 3 gal of acid to 47 gal of water, then the run can be longer. Remember that the safety features shown in Fig. 56 are a must for all new construction where acid use is required. You may specify a safe, dry acid, but you may have no control over what is used at a later date or who uses it.

The type of acid carboy, container, or dilution tank is dependent on the amount of acid used. If calculations show that you will require 8 lb (2 qt) of 60° Baumé sulfuric acid daily, you must use an acid pump capable of pumping that amount at a 50 percent setting. This translates to a capacity of 1 gal/day, or 0.04 gal/h. If you can obtain such a pump readily, there is no problem; however, the best the author could do is a pump operating at 0.08

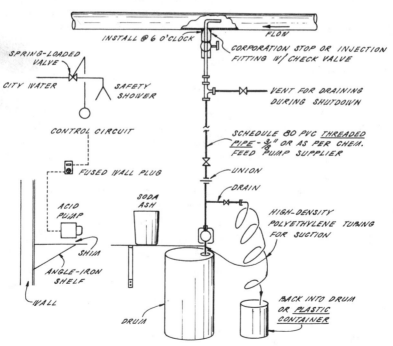

FIGURE 56 Sulfuric acid feed system.

gal/h. In lieu of searching for a low-capacity pump and then hoping it can suck a liquid of close to 2.0 specific gravity, you could choose one of the following alternatives:

1. *Use a percentage timer to reduce acid feed.*

   If the system is large enough, this could be a viable alternative; however, the author has not had good results with using timers to reduce flow. Theoretically it looks good, but in actual practice the results have been very disappointing.

2. *Use diluted acid.*

   The author is presently using this method, and the results are most gratifying. One location mixes 2 gal of 60° Baumé sulfuric acid with 48 gal of water, and the other mixes 3 gal of acid per 47 gal of water. A great deal of heat is generated, but the mixing is done very gradually so the heat poses no problem.

Was the use of a safe, dry acid overlooked? No, not really! When you compute the cost of dry acids, about $1 per pound, and the cost of 60° Baumé sulfuric acid, about $0.10 per pound, then the choice is strongly in favor of sulfuric. If you also consider that 1 lb of sulfuric acid neutralizes twice the amount of a dry acid, then the choice becomes really difficult. You must decide, based on the amount of acid required on a daily basis, the type of feeding equipment available, and the attitude of the operating personnel. More often than not, the last is the deciding factor. Where you have competent personnel and adequate feeding equipment, the choice should be clear-cut.

Can you, should you, connect an acid feed pump as described in order to control the alkalinity? Will it work? To help answer these questions, look at Fig. 57, which illustrates the events taking place.

Since the water meter system is time-controlled via the reset timer, the amount of acid added is always the same, since it is based on a fixed amount of water. Points A, D, and G on the graph indicate the on periods, with B, E, and H being the off periods. Since the acid is added for a set time and is a fixed amount, the pH drop is shown as a straight line until the pump shuts off. At this point the pH could continue to drop, as shown by BC and EF. The amount of drop, or pH dip, is dependent on a number of factors:

1. The concentration of acid used. The stronger the acid, the deeper the dip. This is not too significant as long as the dip does not go below that point at which the corrosion inhibitor is ineffective. This point will vary with the treatment program.

2. Load, or evaporation, on the HX system. As the evaporation, or load, picks up, the alkalinity also increases rapidly. At this point the time period *EG* also compresses since the water meter will trigger the reset timer with greater frequency.

3. Outside contributing factors, contaminants, entering the recirculating water system.

   a. Alkaline dirt picked up by wind will tend to push the pH up as soon as the acid pump is shut off; *HI* shows this trend.

   b. A wind shift could cause combustion exhaust gases, or boiler exhaust, to enter the tower and thus contribute acid to the system. Should this occur, shown at *I*, the pH of the recirculating water system will be pushed lower since the starting point will be lower; *IJ* shows the dip.

Clearly, the use of the water-meter-controlled chemical feed system for acid addition, pH control, is indicated where the natural alkalinity of the water supply is fairly constant. If the alkalinity is not constant, as it is not for the city of Ithaca, NY (see Fig. 58), you must be careful that the cure is not worse than the disease. To base a recommendation on one sample analysis could prove detrimental. However, you are seldom, if ever, afforded the luxury of having the life history of the water supply. A municipality is only obligated to provide potable water and thus performs those tests which seek to prove potability. However, the design engineer, plant engineer, owner, or water treatment supplier is interested in those parame-

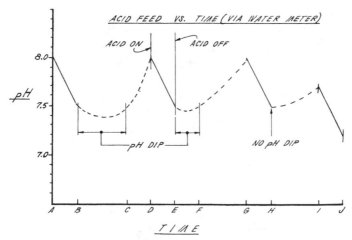

FIGURE 57 Acid feed vs. time via water meter initiation.

ters that will adversely affect the heat-rejecting equipment, for example, pH, alkalinity (P and M), hardness, silica, iron, TDS (by evaporation if possible), chlorides, sulfates, and suspended solids.

In an attempt to control parameters, e.g., in the case in Ithaca, NY, you could install a separate reset timer to control the acid pump and use a dilute acid mix. If tests are run daily, another luxury, you could adjust the feed of the acid over a wide range and maintain very close control of the system alkalinity (pH). Where you are not afforded the luxury of daily tests and the water supply varies, you should consider the installation of a pH controller independent of the chemical feed/bleed system. The decision to install has to be weighed against the costs of not installing it. If a metal cooling tower, an evaporative condenser, becomes encrusted with lime and muriatic acid is used to descale the unit, you can be sure that the life of the tower/evaporative condenser has been drastically reduced! But if by means of proper pH control a safe, dry acid containing inhibitors can be used for descaling, the pH controller has practically paid for itself.

If the decision is made to install a pH controller, then the choice of which one must be addressed, and this is not an easy task. The author has experience with three different manufacturers. The units did do what they were designed to do — control pH — and the least expensive was as good as the most expensive. Some units have a fail-safe function — with a built-in timer — to trigger an alarm and shut off the power to the acid pump in the event that the pH does not reach the off set point within a specified time. To some, this would appear to be a good feature; however, as long as the pH is being controlled within desired parameters, time should not be a governing factor. If the pH is holding at 7.4 and the controller shuts off on fail-safe

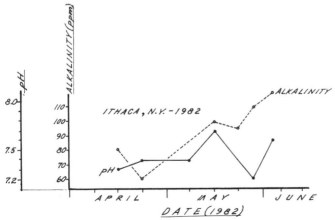

FIGURE 58 Alkalinity vs. timer for Ithaca, NY.

because the pH level has stayed the same for an hour, even though acid is being added constantly, then the alkaline buildup will be very fast. However, to be fair, you have to address the potential for system damage in the event that the pH continues to drop or exceed a set point.

As with a TDS controller, the heart of a pH controller is the pH probe. Once a decision has been made to use a pH controller, you have to address the question of where and how to install the probe. You could play it safe and be guided by the recommendation of the controller manufacturer, hoping that the manufacturer has sufficient experience with the installation of its units in dirty systems such as cooling tower or recirculating water systems. Note that the pH probe is unlike a TDS probe — the pH probe is used to measure voltage, or potential, whereas the TDS probe is used to measure resistance, in ohms. The pH probe, being extremely sensitive to contaminants, must be protected as much as possible, and the controlling circuit must be calibrated more often than with the TDS controller. In a laboratory the pH meter, a measuring device, is standardized daily or before each use by using a standard solution with a pH as close as possible to that being measured. In a pH controller, the probe must be standardized on a regular frequency, ascertained by trial and error.

Standardization of a pH probe should be undertaken following the manufacturer's instructions. Not all probes are alike, and it should not be cleaned in a manner that will adversely affect its performance. The author has standardized instruments by setting the calibration control to make the controller read the pH of the water as determined by another method. For example, we can determine the pH of the water by using pH slides and then setting the controller to that reading. Remember, the easier something is to do, the greater the likelihood that it will be done. If an employee fears breaking the glass probe or altering it, then the probe will never be properly serviced, which could prove detrimental to the system unless the pH is being monitored by independent tests. How can we extend the life of the probe and extend standardization periods? By making certain the probe is exposed to crystal-clean, clear, water! But a recirculating water system is dirty! Figure 59 shows how to install the heart of the pH controller, to extend both probe life and periods of standardization.

Figure 59A shows the probe installed in a by-pass around the circulating pump. If there is more than one pump, we must install the probe across the header to make certain of water flow in case the pumps have to be switched. The cartridge filter, not to be confused with a strainer, is used to filter the water and remove anything that could plug the probe. After the filter we install the probe in a 1-in cross tee. Next we have the option of installing corrosion coupons in two locations, as shown. Since we are in a pressure situation, we need to know the system pressure to avoid exceeding the probe's pressure limitations. Flow is not of paramount importance, but we

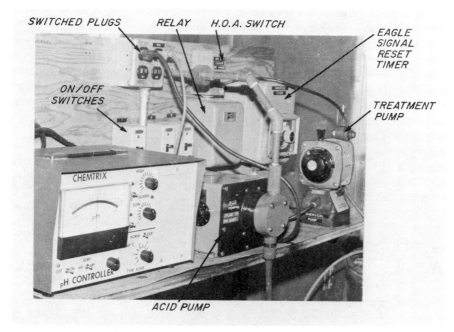

SWITCHED PLUGS   RELAY   H.O.A. SWITCH

EAGLE
SIGNAL
RESET
TIMER

ON/OFF
SWITCHES

TREATMENT
PUMP

ACID PUMP

Water-meter-initiated chemical feed system with pH control.

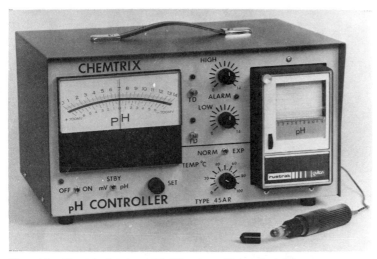

Chemtrix pH controller, type 45. (Courtesy Chemtrix, Inc.)

103

would not want a flow rate set so high that we had to replace cartridges on a daily basis. So a flowmeter would be a nice addition.

Figure 59*B* is essentially the same situation except we are wasting the filtered water. However, this has the advantage of easy probe calibration/ maintenance. The probe is visible, very easily removed, and replaced with minimal, if any, tools. If one does not care for this Rube Goldberg arrangement, then one could consider Fig. 59*C*. Figure 59*B* and *C* may appear to be unprofessional set-ups, but they are, in fact, ideal where outside temperatures preclude freezeups. Both are ideal where it is a 365-day operation with an outdoor pond or indoor open sump.

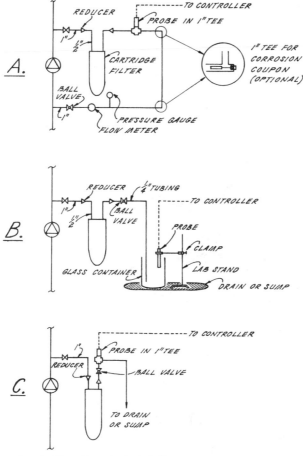

FIGURE 59   pH probe installations.

In the following we assume two things:

1. Acid use is a *must*. You are urged to use caution here and not be misled by knowledgeable or unknowledgeable persons. We say that acid use is mandatory because the alkalinity of the water supply makes it an economic necessity. If you can reduce the cost of water treatment chemicals by 25 percent or more, decrease the load on the environment, and reduce water use as well as insure against the buildup of insulating lime on the HX surfaces, then acid use is a must.

2. The potential user is worried about adding acid to the water.

Thus it is important that the user understand the mechanics of the procedure.

Figure 60*A* illustrates the ideal situation with a functioning pH controller. Since one is constantly fighting increasing alkalinity, the tendency

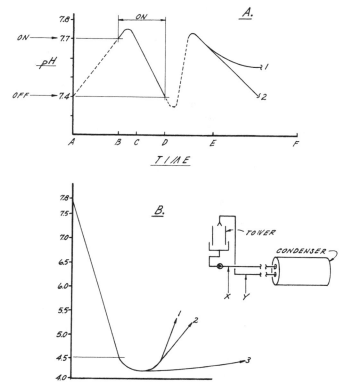

FIGURE 60   Acid addition via pH controller.

is for the pH to climb, as shown by the dotted lines. At a set point the controller activates the acid feed pump, and the pH climb is halted. Note that even though the controller is set at 7.7, the pH will continue to climb until the alkalinity is reduced. This may explain why lime develops on HX equipment even though acid is used. The controller may have cut in, but the pH continues to climb. As soon as the acid being injected equals and exceeds the alkalinity in the system, the pH drops, as shown by the solid line *BCD*. At *D* the controller is inactivated, but the pH curve continues to drop until the incoming alkalinity once again forces the pH up, at *DE*. Curves EF_1 and EF_2 are two possibilities. Curve EF_1 is possible if the acid being injected has a low concentration and the alkalinity buildup is high, e.g., during periods of high evaporation. Curve EF_2 is a straight curve drop characteristic of 60° Baumé sulfuric acid being fed at a high pump setting. With curve EF_2, the dip after the acid ceases will be as deep as *CD*, i.e., point *D*. To prevent high climbs, *BC*, and low dips, one should inject acid of a satisfactory concentration at a proper rate. Here is where the services of a competent water treatment engineer come in handy. The author favors making onsite solutions of sulfuric acid and water; others abhor the practice, citing safety considerations. The matter of safety considerations may or may not be valid depending on the personnel making the solutions. The author will not make any direct recommendations since each plant, like each person, is unique.

In Fig. 60*B* the acid being fed is concentrated, and the pump setting is high. Since we are monitoring at point *X* and injecting acid at *Y*, clearly there is a time lapse between injection point and sampling point. If care were not used, the amount of acid added could be so high as to force the pH down, below safe levels for the system. Yes, the controller will catch the first slug of acid and shut off the pump, but by then the curve will have already started to nose-dive. The recovery rate could be fast as in curve 1, moderate as in curve 2, or agonizingly slow as in curve 3. The solution to a situation such as that shown in Fig. 60*B* is proper engineering!

Corrosion Meters and Controllers Another piece of equipment that could merit consideration is the electronic corrosion meter and/or controller. These units are designed to monitor the corrosion rate and thus alert operating personnel to upsets in the system's chemical balance. Some time ago a unit was available that actuated a controlled sequence when the corrosion rate exceeded a set point. The author has no firsthand experience with it and so cannot provide insight into the pros and cons of that approach.

The reader is cautioned, as in the section on TDS controllers, not to be awed by the term "electronic" or "integrated circuit." Corrosion meters are not children of space age technology but rather mature adults with new clothing. Recall that TDS units function by measuring resistance, in ohms, of a conducting liquid, i.e., a dressed-up ohm meter. Electronic corrosion meters function in much the same way, measuring changes in resistance, in ohms, of an electrode, a piece of metal, as it corrodes. Via this process some units provide instantaneous readings; i.e., a corrosion rate is expressed as conditions stand at the moment of test. Others provide average readings; i.e., time is used as a factor in corrosion rate expression.

Are these electronic instruments cost-effective? Can the expense of installing one be justified? This is most difficult to assess because there are too many factors to take into consideration. For example,

1. Is the recirculating water system used for process or comfort cooling?

2. Under what discharge guidelines is the plant operating?

3. How clean is the system water in terms of the micrometer size of particulates?

4. What is the competency level of operating personnel? Can they understand the figures being read?

5. What is the desired procedure during system upset? Will one dump water or corrosion inhibitor or both?

One factor to consider with these electronic instruments is the interpretation of readings obtained. Figure 61 shows the graph of logged corrosion for a recirculating water system. It is not an actual but rather an illustrative graph to show the difference in instruments. If we had read the instant rate on day 6, it would have shown a rate of 1.5 mils/yr, which is very good by any standard. On day 15 the rate would have been 8.0 mils/yr, which is not too good. An average reading instrument would have provided rates of 2.5 and 6.8 mils/yr. Are these figures significant? If we consider a 6-in pipe with a 0.280-in wall thickness, its life expectancy would have been calculated as 187 years at 1.5 mils/yr, 112 years at 2.5 mils/yr, and 35 years at 8.0 mils/yr. A corrosion coupon in the same water would have provided a corrosion rate of 4.3 mils/yr, 65 years, at the end of a 30-day period. Is the corrosion rate 1.5, 2.5, 4.3, 6.8, or 8.0 mils/yr? Will the system last for 187 or 35 years? At 8.0 mils/yr serious problems will show up as severe leaks, pitting, and general over-all corrosion before the beginning of the 35th year.

Also consider that although instrumentation may give one a good over-all indication of the corrosion trend of clear, clean water, it is not an indicator of actual conditions. Open recirculating water systems, by their very nature, are quite dirty. At the tower they constantly wash the air, gleaning all sorts of particulates, sand, leaves, pollen, filth, etc. out of it, which has a direct bearing on corrosion. System design, dead ends, pockets, etc., also contribute to problems. Thus, although excellent corrosion rates may be obtained with coupons or instrumentation, system failure is still possible if good water treatment practices are not followed. The following points are important:

1. The proliferation of living organisms, bacteria, and algae must be closely controlled. You must maintain a bio-control program and not just add chemicals per se.

2. You must add proper corrosion and scale inhibitors to the system at the correct dosage as determined by the water treatment firm. By "proper" the author implies full knowledge of the system end-use by the water treatment firm. Where a system is used as a closed and open recirculating water system, as in heat recovery, this must be taken into consideration by the water treatment firm and allowances must be made.

3. The pH of the water must be within the range specified by the water treatment firm at all times.

4. Sand filtration must be instituted.

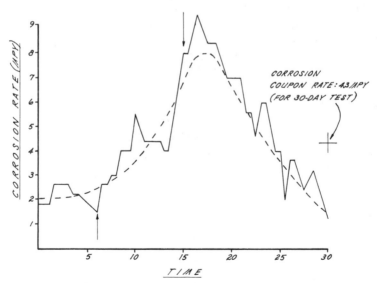

SOLID LINE = "INSTANTANEOUS" READING INSTRUMENT
CURVED DOTTED = "AVERAGE" READING INSTRUMENT
CORROSION COUPON = 30-DAY AVERAGE OF SOLID LINE

FIGURE 61 Corrosion vs. time.

A set of specifications is not included in this section because there are too many variables to contend with. For example:

1. The nature of the water supply (see Chap. 1, "Justification for Water Treatment"). Certainly you would not want to recommend a pH controller where the alkalinity of the water does not warrant it. Even if the alkalinity is high, you still have to contend with calcium and magnesium carbonates (lime), silica, and salt (sodium chloride), to name just a few important parameters.

2. Environmental restraints on bleed-off water.

3. Location of spray water equipment (cooling tower, evaporative condenser, spray pond?) — city, rural, rooftop, parking lot.

4. Use of spray water — process or air conditioning.

5. Environment surrounding spray equipment — hospital, hotel, school, etc.

Should you desire a set of specifications for a particular job, a good place
~~start would be with a consulting firm~~, i.e., consulting engineers who may
...ent consultant. To assist in the
...owing nominal specifications:

...E CAPACITY OF LESS

...lled as shown in the drawings (see

...s than one-half- ($\frac{1}{2}$-) in pipe — PVC,
...solenoid is a one-half- ($\frac{1}{2}$-) in unit.
...then the pipe size should follow suit.

...ll be as follows:

...e size.

...e size.

...lve, ASCO or equivalent, properly
...ndle amount of bleed-off.

...er(s) of the drawing(s) being referred to.

3. *One corporation stop, Precision No. 992 or equivalent.*

4. *One chemical feed pump complete with foot valve, suction and discharge valve, injection valve, and ten (10) feet of tubing capable of withstanding system pressure. (Should additional tubing be required, this shall be provided by the water treatment contractor at no additional charge.) The chemical feed pump shall be a Precision No. 8311-11 or approved equivalent with the motor and gears operating under oil.* Note: *Electronic pumps are not equivalent, nor are piston pumps.*

5. *One heavy-duty fifty- (50-) gallon treatment mixing tank complete with a fiberglass cover.*

This would cover a basic simple chemical feed system. Note that it is not 100 percent complete since you have not stated what will activate the chemical feed pump or solenoid valve. The author recommends activation with the recirculating pump circuit. You can take power from the pump circuit and via a suitably sized relay (do not forget to state who supplies this relay) supply the 110 V for the chemical feed pump and solenoid valve. Of course, you could specify that the activator shall be a tab-type timer with a 24-h dial rotation. Should you lean toward a TDS controller, then Figs. 47, 48, and 49 should be reviewed and the specifications altered to accommodate that decision:

6. *One TDS controller, piped as shown in the drawings (see Fig. X),* Morr control model JA-1 with a three-quarter- ($\frac{3}{4}$-) in NPT, PVC, flow tee with a three-quarter- ($\frac{3}{4}$-) in PVC electrode capable of withstanding 100 psig.*

7. *One cartridge filter complete with one (1) year's supply of filter cartridges (26 cartridges) with a service flow rate of six (6) gal/min and supplying water with particulates down to five (5) μm. The filter material shall be white cellulose fibers with graduated density. The unit shall be a Teel 1P747, Cuno, or approved equivalent.*

8. *One flowmeter, one-half- ($\frac{1}{2}$-) in RCM, SK, or approved equivalent.*

9. *All wiring shall be done by electrical contractor.*

Recall that this chemical feed system is a basic system for units of under 75 tons. What has been outlined are the *minimum* items necessary, piped in a manner that will not bring down criticism on the design engineer, to make a system work with some degree of certainty. Anything less, e.g., a smaller-capacity cartridge filter or lack of a flowmeter, than that specified could reflect poorly on the design engineer. The items shown in Figs. 48

---

* In an actual guideline, insert the correct number(s) of the drawing(s) being referred to. In this book, we are referring to Fig. 47.

and 49 may be required depending on local water conditions and whether the arguments outlined in the text are valid. The reset timers shown are Eagle Signal HP53s and are wired as shown in Fig. 51.

Equipment with over 75 tons of cooling capacity should have a water-meter-activated chemical feed system. This is not to say that one should fail to consider a water meter system for systems under 75 tons. Rather, at this point, one should be considering it not as an option but as the only route to take. Here are sample specifications:

EQUIPMENT WITH A HEAT-EXCHANGE CAPACITY OF 75 TONS AND GREATER — WATER METER CHEMICAL FEED SYSTEM

The water-meter-activated chemical feed system shall be installed as shown in the drawings (see Fig. X)* and shall consist of the following:

1. *One electric-contact head water meter with totalizer sized for system tonnage [list] by water treatment contractor, Carlon J.S.J. industrial water meter or approved equivalent. Body shall be bronze, register shall be hermetically sealed, and drive shall be magnetic.*

2. *One Eagle Signal HP53 reset timer complete with HN364 case. The time duration of the timer shall be as sized by the water treatment contractor.*

3. *One hand-off-auto switch, Furnas or approved equivalent, complete with enclosure.*

4. *One 25-A relay, complete with enclosure, Furnas or approved equivalent.*

5. *One solenoid bleed-off valve of no less than one-half ($\frac{1}{2}$) in and sized by the water treatment contractor for system bleed-off; ASCO or approved equivalent.*

6. *One flowmeter, sized by the water treatment contractor for system bleed-off, SK, RCM, or approved equivalent.*

7. *One chemical feed pump suitable for system pressure and compatible with chemicals to be used. If the chemical feed pump does not have its motor operating under oil, e.g., Precision No. 8311-11 pump, then its motor must be a high-efficiency unit and not operate hot. The chemical feed pump must come with foot valve, suction and discharge valve, and injection fitting. The pump shall be Precision No. 8311-11, mRoy No. 37R-100, Neptune 500-A-N, or approved equivalent.*

8. *One corporation stop, Precision No. 992 or approved equivalent.*

---

* In an actual guideline, insert the correct number(s) of the drawing(s) being referred to. In this book, we are referring to Fig. 55.

9. If acid is to be used for alkalinity control, one chemical feed pump constructed of materials suitable for use with 60° Baumé sulfuric acid.* This pump must be of same manufacture and type as the water treatment chemical feed pump — Precision, mRoy, Neptune, or approved equivalent. The installation of this pump must be done per drawings (see Fig. X).† It will be the responsibility of the water treatment contractor to provide supervision during installation and to supply suction tubing, injection valve, and proper injection fitting. Further, it shall be the responsibility of the water treatment contractor to provide a source of PVC pipe and fittings for the acid pump. The mechanical contractor shall purchase the necessary fittings per the advice of the water treatment contractor.

10. One by-pass feeder, minimum two- (2-) gallon capacity, piped across the circulating pump or header as shown in drawings.

These nominal specifications should serve as a guide and should be tailor-fit to any particular job based on local conditions. The bias in the author's equipment selection is due to a healthy experience with a particular product or item.

---

* NOTE: If sulfuric acid is used, the author highly recommends the elimination of the HOA switch and the installation of a momentary-contact switch, Furnas or equivalent, across terminals 1 and 4. The HOA switch should serve as an on/off switch with no hand, or manual, option.

† In an actual guideline, insert the correct number(s) of the drawing(s) being referred to. In this book, we are referring to Fig. 56.

FOUR

Algaecides/Biocides

THE PROBLEM

With the advent of *Legionnaire's Disease* (LD), operating personnel have been paying greater attention to the control of bacteria and other living organisms in their recirculating water systems. With the advent of bio-degradable products, the control of these organisms has been difficult, and it is getting more so. Here the term "bio-degradable" means the breakdown of a product by living organisms in the environment. Bio-degradation does not begin when water, such as bleed-off, is discharged into the environment. Unfortunately, for the user, it begins as soon as the product leaves the drum — if it has not already begun. Since a bio-degradable product is being added to a system and since a recirculating water system is basically an air washer, steps must be taken to insure that living organisms in the system do not multiply unchecked. Let us look at some living organisms as groups and see how they affect open recirculating water systems. The following limited and general coverage is meant to provide an insight into problem areas.

Algae

Growths of algae can reduce tower efficiency by reducing its surface area and thus hindering evaporation. Plugging of strainers is a significant problem when plugs of algae slough off the slats or walls of the tower. Under-surface corrosion or delignification of wood at an active growth site and at crevices is a major concern.

Slime Growth

A slippery mass that allows undersurface corrosion to proceed unabated, since it forms a barrier between the subsurface metal and corrosion inhibitor, is a slime growth. There is also the possibility of acid secretion, by the organisms, causing accelerated corrosion.

Fungi

Incapable of producing their own food, fungi are dependent on outside food sources, e.g., killed bacteria, organic debris gleaned from the atmosphere, etc. Some in this group are capable of utilizing wood as a nutrient source and thus will attack tower lumber.

Bacteria

This group has received a great deal of attention, owing not to equipment operation but to health considerations. But let us consider both areas of concern.

Equipment If sulfuric acid is used for alkalinity control and/or the water contains sulfates, the proliferation of acid-producing bacteria is possible. The author has run across this problem on many occasions, and it has proved difficult to solve. It is easy, of course, to appreciate the ramifications of operating a system under very low pH — from 2.0 to 4.0 — for any length of time.

Another group that causes equipment problems is the iron bacteria. This group utilizes iron, from the system, in its metabolism and can thus shorten equipment life. Bacteria can form slime in the system and thus reduce heat transfer and efficiency. Other bacterial groups can produce foul orders, foaming, and delignification of wood, all undesirable.

Health The impact of LD on HVAC personnel has been significant, alerting operators to the potential hazards involved in treatment neglect. Let us consider possible mechanisms involved in airborne health hazards associated with open recirculating water systems and then turn our attention to control to prevent problems.

A normal human body temperature is about 98.6°F (37°C), within the operating range, 85 to 105°F (29.4 to 40.6°C), of an open recirculating water system. Thus conceivably we could assume that if something can grow, or metabolize, in the system or environs, the potential for its growing in the human lung is present. Now let us consider that bio-degradable products are being added to the system, thus increasing the bacterial population density. Further, as for the -cides routinely added for the control of living organisms, none are capable of destroying 100 percent of all organisms within the system. Thus it stands to reason that a percentage of the organisms will develop an immunity to whatever program is being used, not be affected by the control program at all, or utilize the bio-control agent in their metabolism since it is bio-degradable.

The only possible exception could be the use of chlorine; however, even a vigorous chlorine program is incapable of achieving system sterilization, and even if it were possible, the HVAC equipment could not withstand the high corrosion rate induced by vigorous chlorine use. For an in-depth, but not all-inclusive, insight into bacterial problems, from a health standpoint, study the work done by the author (see Ref. 6). The study was an attempt to answer the question, What is a high bacterial count? It concluded with astonishing observations, leading to recommendations for the user.

How does one get sick? That is a question the author has had to field more than once. In the aforementioned study, the normal bacterial count was in excess of 100,000 organisms per milliliter, with 30 mL making up 1 oz. The count, of course, did not include protozoans, algae, fungi, and those organisms incapable of utilizing the medium in their metabolism. If a 100-ton tower, open recirculating water system operating at 100 percent load evaporates about 3.0 gal/min (11,355 ml/min), that translates to $11,355 \times 100,000 = 1.1 \times 10^9$ living organisms dumped into the atmosphere, downwind of the tower or fan, each minute. Now consider a cooling system that has been under poor care, neglected, and shut down for economic reasons for some time. *Anyone downwind of such a system on startup would literally be enveloped by bacteria, inhaling all sorts of living organisms.* Is this bad? If a significant percentage of the organisms are capable of hemolytic activity, then the potential for adverse health effects is present. If this activity is low or absent, this potential is lowered, but not altogether absent since organisms exist that may not exhibit hemolytic activity but are harmful. Further, it should be noted that LD (actually it should be LD's since more than one organism is involved) cannot be detected with Nutrient Agar or Trypticase Soy Agar with 10 percent sheep blood but it can cause illness. There is now a prepared agar that may prove of value to anyone who wants to test for LD. The author, however, has had no hands-on experience with the medium or procedure as of publication date so is hesitant to comment further. If found to be of value once it has been tested, information will be made public to the reader. It would be interesting to ascertain the number of HVAC personnel who stay home with a cold or flu a day or so after servicing a poorly maintained system.

CONTROLLING THE PROBLEM

Now that we have some insight into problems caused by living organisms in recirculating water systems, we can appreciate the importance of their control. At present, the technique consists of adding a proprietary-cide to the system on a regular or irregular basis. The success, or failure, of this approach depends on the bacterial flora of the system, type of -cide program used, and frequency of application. Since this is not meant to be a technical treatise or textbook on bacteriology, we concern ourselves only with the application of the proprietary-cide program.

The simplest approach is to apply the -cide on a regular basis as recommended by its formulator. With an open pan/sump cooling system that is easily reached, this does not pose a problem. However, with closed-circuit coolers it does. As noted earlier, the author believes in the maxim "The greater the difficulty or effort required to accomplish a simple task, the greater likelihood it will not get done." This does not speak badly of maintenance personnel; it only defines human nature. The objective, then, is to make the task as effortless as possible.

When the tower is not readily accessible but the recirculating pump is, one approach is to use a hand pump as shown in Fig. 13. The pump is chained to a location, along with a 2-gal steel bucket, a schedule is set up, and chemicals are injected at that point. This approach does not require the maintenance personnel to brave the rain, climb ladders, shut off pumps, open doors, etc. All that is required is to pour the proper amount of -cide into the bucket, pump it in, and follow with some clear water.

Another simple approach, excellent for easily accessible closed-circuit coolers, involves the installation of a funnel on the suction side of the pump, as shown in Fig. 62. It is important that the valve be higher than the water level in the sump. With this set-up, the fan or circulating pump does not have to be turned off. The -cide is poured into the funnel, and the valve is opened.

Both the hand pump and funnel arrangement require someone to physically do something, which is fine if personnel are available. However, what about locations lacking personnel altogether, such as those where an HVAC firm visits for startup and shutdown only, with perhaps an occasional visit to correct a problem? From what the author has seen of these locations, they seem to be bacterial time bombs. It is only a matter of time before someone gets LD from one of these systems and litigation demonstrates the importance of proper maintenance. For these systems and where automation is desired, the set-up shown in Fig. 63 is recommended. The Tork 8007V timer can be set to activate the chemical feed pump, properly sized by the formulator, and close the normally open (NO) bleed-off solenoid valve for 3 h or more as recommended by the -cide formulator. Where added insurance is desired, another pump could be connected to the timer and the -cides could be alternated. Some very nice prepackaged

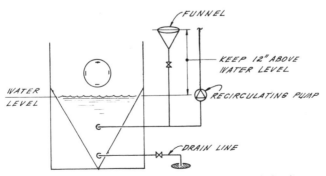

FIGURE 62 Chemical addition via funnel for closed-circuit cooler.

systems are on the market that perform essentially the same task; the choice is left to the user. The biggest disadvantage of using the latter is replacement parts; with an off-the-shelf item, parts are easy to acquire.

Another approach involves the use of chlorine on a monthly basis at high initial dosages. What the user may fail to understand, however, is that the effect is short-lived, with the bacterial population recovering by the fourth day after addition. This is the result of the dilution effect of make-up water and the venting action of the fan. Further, since *hypochlorous acid* is the active agent, the killing action is favored at a low-pH range, far below the actual system pH. At a pH of 7.5, the normal low end of most open recirculating water systems, the efficacy drops to 50 to 60 percent. At higher pH, the efficacy drops to less than 25 percent. So, although a test kit may show a high chlorine content, the killing potential may not be as great as expected. For shock dosages to be effective, they must be controlled and administered at the correct amount and maintained at the correct pH.

To control the amount of dry chlorine or bromine compound added to a system, the use of by-pass feeders, as shown in Fig. 29, is recommended. If there is a daily testing program, the system is kept clean, the residual is kept below 2.0 ppm, and conscientious personnel are available, this method could prove satisfactory. If bromine is the agent of choice, it should not be the sole agent used. The author has seen systems where bromine was being maintained at 2 to 3 ppm, and was not favorably impressed. The

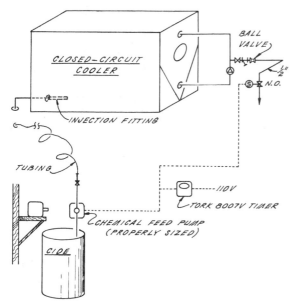

FIGURE 63 Algaecide addition via timer.

reader should also consider that these compounds, being oxidizing agents, will shorten life appreciably. Further, note that these compounds cannot differentiate among metal, wood, organic corrosion inhibitors, airborne organic debris, or bacteria — all will increase consumption of bromine.

An important point to remember is that chlorine and bromine compounds will alter the *chloride* test so that it will be impossible to ascertain true cycles of concentration if chlorides are used as a measuring parameter. Although a 1.0-ppm compound may be meticulously maintained in the system, there is no way of ascertaining the amount consumed and contributing to the chloride reading. Further, some test kits are adversely affected by chlorine and bromine in the test sample, giving odd readings if this is not compensated for.

The use of liquid chlorine, along with control equipment, has been suggested to the author. However, the costs of installation, maintenance, and chlorine consumption were negating factors. Liquid chlorine is not a compound that can be added without adequate controls; pH and chlorine controllers must be used.

As can be seen, the control of living organisms in a recirculating water system is not an easy task, nor one that should be neglected. The author recommends you follow, with exactness, the recommendations of the proprietary-cide manufacturer for the following reasons:

1. The formulator knows the exact amount of active ingredient in the

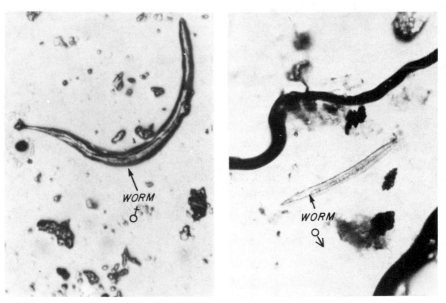

Nemathelminthes found in cooling tower using conventional -cide program.

product. To use more violates Environmental Protection Agency (EPA) laws; to use less invites problems.

2. In the event of litigation, you could be asked to demonstrate, via log sheets and purchases, that an adequate, up-to-date biocontrol program was being followed.

3. An ill employee is not very productive.

Should you desire to start a bacterial testing program, which is not an altogether bad idea if the cooling tower is located close to a populated area, consult someone with bacteriological experience. The author has done extensive testing[6] and found the following methods provided valuable information.

Cultivating bacteria with nutrient agar. Nutrient agar is a general microbiological culture medium that can be used with or without blood for the cultivation of bacteria. It will not cultivate all bacteria, but what does grow in it will be useful in providing an indication of the bacterial population density in the recirculating water system. Employing this medium will also be helpful in determining the usefulness of a particular biocide program. If the biocide is effective, its addition should lower the density of the bacterial population in the recirculating water system.

Determining hemolytic activity trypticase soy agar with 10 percent sheep's blood. Besides not being satisfied with the use of nutrient agar, the author lost interest in knowing what the bacterial population was

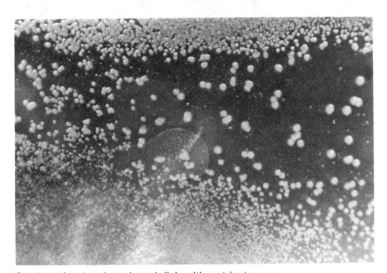

Section of an incubated petri dish with nutrient agar.

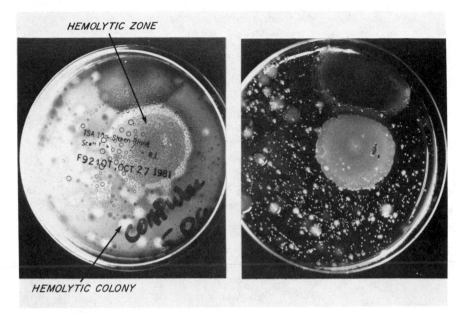

TSA agar showing hemolytic zones.

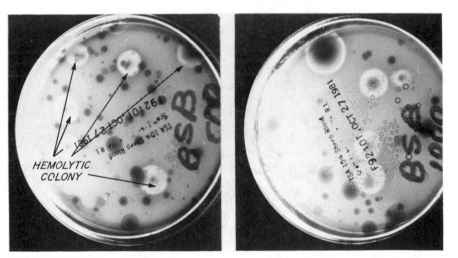

TSA dishes at different dilutions.

for one project after learning that the total plate count was meaningless for the results needed.[6] So this agar was substituted to assess the hemolytic activity of bacteria present. The results were so good that it has become the author's agar of choice for ascertaining biocide effectiveness. It can be prepared with human blood, but to do so will be expensive and potentially harmful to the person doing the testing. Sheep's blood is adequate for determining hemolytic activity.

You will require, in addition to the above, an incubator capable of incubating at 98.6°F (37°C) in a saturated atmosphere, colony plate counter, sterile water in 1000- and 500-mL disposable containers, and an adequate supply of disposable pipettes. The agar should be purchased in a prepared state for ease of use and disposal, especially Trypticase soy agar with 10 percent sheep's blood. One never knows what one may be cultivating with this medium. The author disposed of his unopened, used, plates by first immersing them in a 5-gal container with a solution of 1 gal of sodium hypochlorite (5 percent household bleach) and 4 gal of water. The next day the plates were thrown out with the garbage.

Most of, if not all, the material necessary for setting up the above can be obtained from

Fisher Scientific
711 Forbes Avenue
Pittsburgh, PA 15219
(412) 784-2600

Bacterial colony counter. (Courtesy Hellige, Inc.)

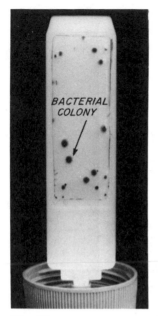 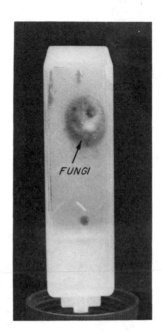

Dip slides.

The author does not recommend the use of the dip slides, touted by some water treatment companies, because they will not provide the user with any meaningful information. The author has found no correlation between the results with dip slides and those obtained by using standard procedures. Further, you will be interested to know, from a health standpoint, whether the biocide program being used is effective. *Hemolytic activity would appear to indicate unhealthful conditions,* whereas a high count on a dip slide is meaningless.

The author would like to recommend that during spring startup and fall shutdown of a system it be thoroughly cleaned and sanitized. The cleaning procedure may be a modification of that suggested in Chap. 2, "Closed Systems." The final flushing should be done with a sanitizing agent containing chlorine. The chlorine should be kept at 50 + ppm for 8 h, with the pH kept close to 6.0. During this time all parts of the system should be sprayed and scrubbed, if possible, with the sanitizing agent solution. Although this procedure will not guarantee 100 percent protection against bacterial problems, it will certainly minimize them.

FIVE

Steam Boilers

Boilers, like open recirculating water systems, are prone to corrosion, pitting, and scale formation if preventive steps are not taken. Pitting, or accelerated pinpoint corrosion, is the greatest enemy of boiler tubes. One failure, or pit, may necessitate replacement of a tube, or tubes, at significant expense and at the most inopportune moment. To prevent problems, at the planning stage the system should be designed for longevity by providing the means for a good water treatment program. Since a treatment program is dependent on boiler pressure and use, we discuss boilers in terms of their pressure range — low (under 15 psig), medium (below 600 psig), and high to supercritical (above 600 psig). These ranges are not absolute — some engineers even consider 200 psig as high pressure — but they will serve to define the approach.

WATER ANALYSIS

As with any system using water, the primary analytical tool, the water analysis, must present an ordered list of parameters, to provide those bidding on a system a means of assessment. It is impossible to offer advice or a quote on a system without a water analysis to review. Do not assume that

1. Water does not change.

2. All water treatment companies quoting on the job have analyses of the water available.

It would be advantageous to tabulate the analytical results of the water expected to be the supply for the job site. The following table should serve as a guide:

Parameter	Values	
	Raw	Soft
pH	———	———
P Alkalinity, ppm as $CaCO_3$	———	———
M alkalinity, ppm as $CaCO_3$	———	———
Hardness, ppm as $CaCO_3$	———	———
Silica, ppm as SiO_2	———	———
Chlorides, ppm as NaCl	———	———
Conductance, micromhos	———	———
TDS by meter, ppm	———	———
TDS by weight, ppm	———	———

The analyses can be supplied by those water treatment firms that have adequate laboratory facilities. A 1-gal sample of the raw water should be sufficient for the water treatment company's analysis since only about 1500 mL is needed to provide all the information listed. The other 2500 mL can be run through appropriate water treatment cartridges to obtain a proper sample for analysis. Excellent disposable water treatment cartridges are available commercially, and a basic water softener can be readily constructed (see Fig. 64). The resin, at a standard 30,000 grains per cubic foot (gr/ft^3) of softening capacity, can be purchased from local firms that service water softeners. The unit, as shown, will have a capacity of 0.17

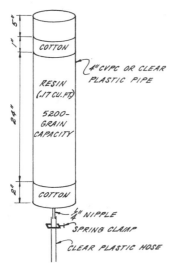

FIGURE 64 Simple water softener.

ft^3 which should provide about 5000 gr of softening capacity before the resin is discarded or regenerated. Using water as supplied to the city of Syracuse, NY, which is about 8.0 gr hard, one should be able to obtain about 600 gal of soft water.

Another possible source of information are those local — to job site and using the same water supply — firms with significant power plants. Most plant engineers are very cooperative and would be happy to provide the necessary information from their boiler room logs.

BOILER SYSTEMS AND WATER TREATMENT PROGRAMS

Let us now address each of the three types of boiler systems and see how we can specify a water treatment program that will provide for equipment longevity.

Low-Pressure Boilers

We begin with low-pressure units, and first we specify guidelines for a treatment program:

LOW-PRESSURE STEAM-HEATING BOILERS UNDER 15 psig

1. *Existing system (renovation)*

   a. *Existing boiler with a new steam distribution system — cleaning*

   The existing boiler(s) shall be inspected by the water treatment contractor, who shall recommend the proper chemical compound and cleaning procedure, if required, to clean the waterside of any and all incrustations. The water treatment contractor need not supply the cleaning agent but must provide the mechanical contractor with the name and source of the recommended cleaning agent. During the cleaning procedure all condensate is to be wasted to drain.

   On completion of the cleaning procedure, the water treatment contractor shall inspect the waterside and ascertain the extent of cleaning. If conditions appear satisfactory, the mechanical contractor shall then contact the clerk of the works for a final inspection. Should the clerk of the works find unsatisfactory conditions, the cleaning process shall be repeated, at no additional cost. If conditions are satisfactory, the boiler shall be filled and treatment added the same day.

   b. *New boiler with existing steam distribution system — cleaning*

   The boiler(s), waterside, shall be cleaned of any and all manner of oil, mill scale, dirt, and debris due to construction prior to startup. The chemical to be used shall be supplied by the water treatment contractor and shall meet all local, state, and federal pollution laws, rules, and regulations pertaining to the discharge of spent cleaning agent. Compliance shall be the sole responsibility of the water treatment contractor. The mechanical contractor shall follow all written instructions supplied by the water treatment contractor to the letter — no deviations allowed.

Upon completion of the cleaning procedure, the mechanical contractor shall have the boiler inspected by the clerk of the works. If conditions appear satisfactory, the boiler(s) shall be filled and treatment added the same day.

It is to be noted that shortly after startup conditions may be such as to cause high treatment consumption owing to the increased blowdown requirement caused by high condensate loss. The mechanical contractor shall include sufficient funds in the bid to repair the return-line system. This includes steam traps, unions, elbows, and piping. Should the problem rest with a defective heat exchanger, the mechanical contractor or water treatment contractor shall bring it to the attention of the clerk of the works. The clerk of the works shall then take up the matter with the owner, who shall take the necessary steps to rectify the condition. The cost of increased chemical use shall be borne by the owner until conditions are corrected.

2. *New construction — cleaning*

   The boiler(s), waterside, shall be cleaned of any and all manner of oil, mill scale, dirt, and debris due to construction prior to startup. The cleaning chemicals shall be supplied by the water treatment contractor and meet all local, state, and federal laws, rules, and regulations pertaining to the discharge of spent cleaning solution. Compliance shall be the sole responsibility of the water treatment contractor. The mechanical contractor shall follow all written instructions supplied by the water treatment contractor to the letter — no deviations allowed.

   During the cleaning process, all condensate shall be piped to waste. If steam was not generated during the cleaning procedure, then after completion of cleaning and treatment addition the boiler(s) shall be brought to operating pressure and the condensate wasted for a period of not less than one (1) hour. At completion of the allotted time period, the boiler(s) shall be brought down, cooled, flushed, and refilled. The condensate tank shall likewise be flushed.

3. *Water treatment*

   *Upon completion of the cleaning process, each boiler shall receive an adequate amount of treatment to guarantee the following for contract duration.**

---

* NOTE: The contract period is normally for 1 year from date of startup, even though the boilers may operate for only 6 months.

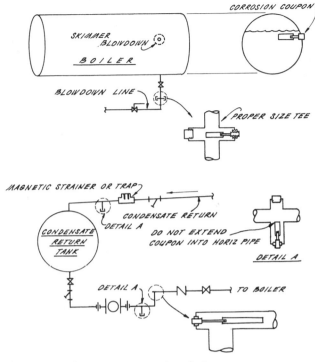

FIGURE 65 Corrosion coupon installation.

  a. Scale

    The chemicals used, in conjunction with the blowdown program estab-lished by the water treatment contractor, shall prevent the formation of scale or incrustations of any kind. Should scale or incrustations become evident, during operation or at contract's end as the unit is opened for inspection, they shall be cleaned at the expense of the water treatment contractor, assuming the owner complied with instructions as verified by the boiler water treatment log.

  b. Corrosion/pitting

    Chemicals supplied shall control corrosion in boiler(s) internals and return lines to less than five (5) mils/yr,* as determined by NACE methods or corrosion coupons.† Pitting, pinpoint attack, will not be tolerated, and immediate, effective means shall be used to rectify if coupons detect this process.

---

* At this rate, one can expect a life of 20 + years for a 0.110-in-thick boiler tube.

† NOTE: Make provisions for installing the corrosion coupons; see Fig. 65.

c. *Suspended solids*

The treatment supplied and recommended blowdown procedure shall control this parameter so as to prevent operational difficulties due to suspended solids.

d. *Total dissolved solids (TDS)*

This parameter shall be maintained below that level associated with priming, foaming, and carryover. The water treatment contractor shall provide a test procedure to control this parameter. Under no circumstances is conductance to be converted to a TDS figure and reported as such. Conductance shall be reported in micromhos or directly in ohms.

e. *Testing equipment*

The water treatment contractor shall supply the necessary test kits to ascertain:
(1) Concentration of corrosion inhibitor
(2) Hardness, using a test kit with EDTA as the reagent
(3) Chlorides, ppm as NaCl
(4) pH via a pH slide operating within 7.2 to 8.8, color wheel, or pH meter
(5) Any other test kit deemed necessary by the water treatment contractor

f. *Service*

The water treatment contractor shall, at startup, instruct the owner's representative or staff on all test procedures, blowdown recommendations, and maintenance of proper records. Further, during the operating season the water treatment contractor shall make monthly service calls to assess the status of the water treatment program. The water treatment contractor shall also be available within 48 hours to assist with any problems which may occur.

All chemicals supplied shall meet federal, state, and local environmental laws, rules, and regulations. OSHA safety data sheets shall be supplied with submittals.

As can be seen, the specifications do not limit the choice of water treatment chemicals, nor do they provide a carte blanche. The choice, as well as the onus, is left up to the experts employed by the water treatment company. However, in all fairness, since we expect a certain performance, we should also provide the mechanical means to achieve it. The ideal chemical feed system for low-pressure heating boilers is shown in Fig. 66.

The steam traps at the top of the boilers are to prevent boiler flooding in multiple-boiler installations during low-demand periods (see Fig. 10). This will reduce make-up water since draining a flooded unit will not be necessary because the units will not flood.

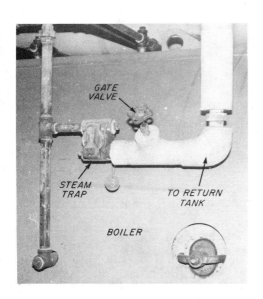

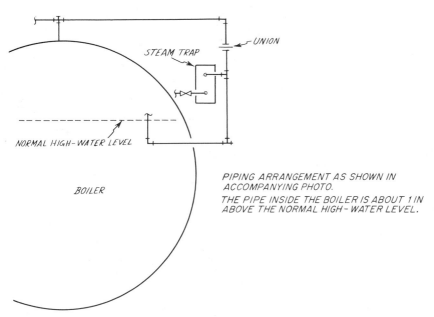

Steam trap off boiler side to prevent boiler flooding.

Since we are dealing with heating boilers, which should have almost 100 percent return condensate, the use of chemical feed pumps is not indicated. The simple feeders shown are more than adequate for low-pressure heating water treatment formulas. Should the water treatment contractor decide to supply a sulfite-based product, then perhaps a chemical feed pump and tank with proper means of control may be necessary. However, the water treatment contractor should supply the means necessary to add the treatment to the boilers. The design engineer should not be expected to anticipate every nuance or requirement of every water treatment contractor. It should be sufficient to provide for an adequate, simple program such as chromate-, nitrite-, or molybdate-based formula. The author realizes that chromates are frowned upon and that nitrites, when used in conjunction with amines, could pose health problems where steam is allowed to escape. However, all good water treatment companies can provide an adequate, simple program. Should the water treatment contractor deem a chemical feed pump necessary or should the program require one, it must be supplied as part of the quotation to the mechanical contractor. The owner should not be expected to absorb the cost of a chemical feed pump and tank as a surprise package.

The water softener, which is not required if cistern water is used, serves as insurance against condensate loss. The size of the unit should be based on the flow required to support the boiler in the event of a catastrophic return-line failure rather than on anticipated condensate loss.

The magnetic strainer, or trap, as in Fig. 24, on the main return line removes rust particulates before they enter the condensate tank and thus boiler(s). It is important to remove these particulates because they can lead to tube failures, via pitting, where the particulates build up. Further, the rust can also form a sticky sludge detrimental to boiler efficiency. Do not

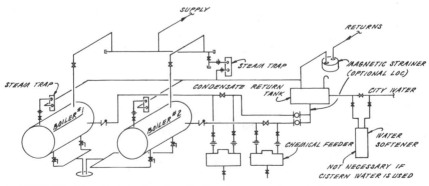

FIGURE 66   Chemical feed system for low-pressure steam boilers.

overlook the impact of abrasive particulates on the mechanical seals or packing of the boiler feed pumps.

Where a low-pressure heating boiler is used as a process unit, e.g., for humidification, we must make certain allowances. We can use the same basic specifications as for the heating boiler, with the only difference being the make-up water requirements. The chemical treatment program will vary, but the specifications do not spell out what chemicals or formulas to use. Performance rather than product is specified. The specifications do not hinder a water treatment contractor from setting up whatever program is deemed necessary to obtain the specified ends. As for a chemical feed system, the by-pass feeder would not be adequate for a humidification boiler. Each boiler must have, in lieu of the by-pass feeder, its own chemical feed pump and tank, if necessary, interconnected with the boiler controls or a tab-type timer. Whenever the boiler comes on steam, its chemical feed pump must become operational.

Since we are addressing a situation where make-up water will be a significant variable as the load varies, the primary concern with a humidi-

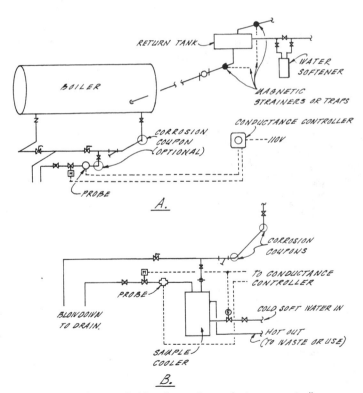

FIGURE 67 Automatic blowdown via conductance controller.

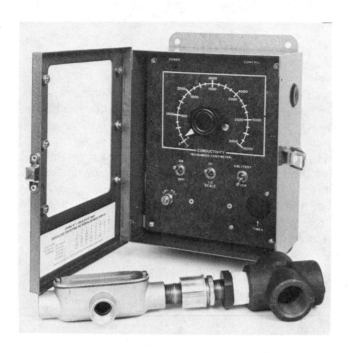

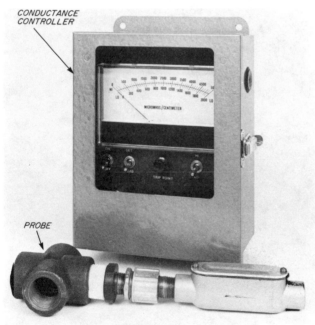

(a) Morr conductance controller system 1-A; (b) Morr conductance controller system 1. (Both photos courtesy Morr Control, Inc.)

fication boiler is adequate blowdown! Obviously blowdown must keep pace with the load or keep ahead of it; however, rarely is this the case. The boiler may get blown down according to a set schedule, perhaps based on results of chemical tests, whether it needs it or not. Since boiler longevity is dependent on care received, provisions must be made at the design stage to insure that the boiler can be blown down. Installing gate valves under boilers, where they are inaccessible, does nothing to insure boilers will get the necessary blowdown. Quick-opening blowdown valves that are easily accessible should be utilized for this type of operation.

Another possibility that has merit is an automatic blowdown system. This can be accomplished via the use of automatic blowdown controllers or timer-actuated motorized blowdown valves. The author is hesitant to suggest that automatic blowdown should be the owner's answer to a sloppy boiler room operation. Automatic equipment should not be used to replace the human element, but rather to strengthen its weaknesses. It is obvious that unless there is a 24-h crew conscientiously attending the boiler, it will not get the ideal attention required for optimal efficiency. If it is used correctly and its operation is understood, the automatic equipment can take over from the operating crew when necessary. But it is not to be used in lieu of a boiler attendant! With this understanding, let us look at Fig. 67 and see how to pipe an automatic blowdown system.

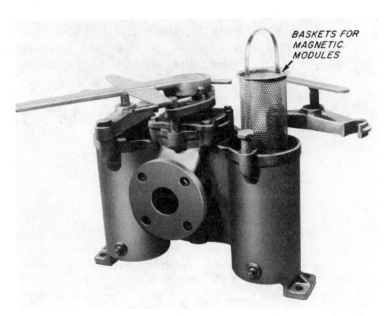

Hayward model 50 duplex strainer. (Courtesy Hayward Manufacturing Company.)

First, and foremost, **the author does not recommend that a conductance controller be used where the make-up water is not softened.** It is imperative to minimize sludge formation in the boiler, and a properly maintained water softener will do just that.

The next item to consider, perhaps as important as the water softener, is the magnetic strainers, or traps. They can be installed in the return line or before the boiler feed pumps. The traps will remove iron particulates and thus prevent their interfering with the blowdown controller.

The probe of the blowdown controller is shown installed on the bottom blowdown, front or rear. It can also be installed at the skimmer level; however, not all boilers have tappings for skimmer blowdown, whereas all have bottom blowdown connections.

If local regulations prohibit hot water from being drained into the sewer system, you can install a sample cooler as shown in Fig. 67*B*. The blowdown controller, operating the motorized valve on a time basis, also actuates the solenoid valve and thus cools the water. The cooled boiler water can be sampled during the activation period, or the controller can be manually activated and tested for compliance with set parameters. Bottom blowdown water is the preferred source for sampling since it will not be contaminated with condensate as would water taken from the gauge glass.

In operation the blowdown controller, via an internal timer, actuates the motorized valve and allows the boiler water to be sampled by the probe. If the conductance of the boiler water, at that pH and temperature, is within the desired parameters, the controller shuts off the motorized valve and repeats the process at the next time period. But should the conductance exceed the set parameters, the controller will keep the motorized valve open until the conductance comes down to the set point. The throttling valve, with micrometer dial and pointer, is set very low so as not to affect boiler operation in case the controller malfunctions and holds the motorized valve open. You may wish to utilize the following specifications:

AUTOMATIC BOILER BLOWDOWN

A blowdown controller shall be installed as shown in drawings. The water treatment contractor shall be responsible for supplying a complete system as follows:

1. *One timer-actuated automatic conductance controller using one hundred and ten (110) V ac with an output current of ten (10) amps. The enclosure shall be moistureproof and the electronics easily removed for service by the manufacturer. The manufacturer shall have a factory exchange program to insure that the controller is not off the line longer than forty-eight (48) hours or the time it takes for a replacement unit to arrive from factory via first-class mail.*

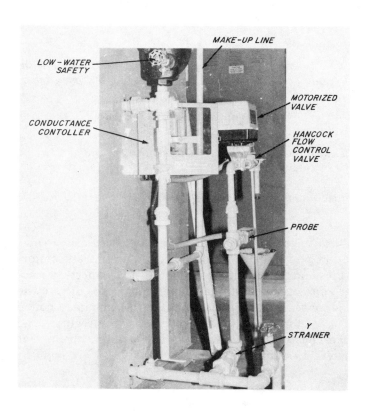

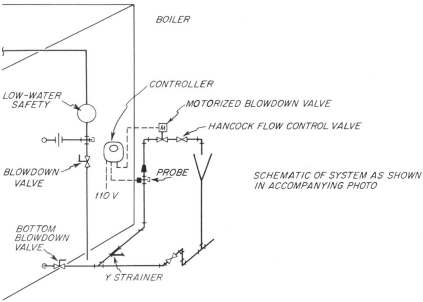

Automatic blowdown controller for steam boilers.

2. *The probe body with one- (1-) in normal pipe thread (NPT), shall withstand at least two-hundred- (200-) psig steam, and the electrode shall be stainless steel and easily cleaned.*

3. *The motorized valve shall withstand at least two-hundred- (200-) psig steam, and the seat shall be of 315 stainless steel.*

4. *The throttling valve shall be a globe type with a micrometer dial and shall be designed for intended use.*

The design conductance controller shall be a Morr boiler master system 1-250 or equivalent.

With the advent of the energy crunch and the increased use of computers, humidification has been employed more often for human comfort at lower temperatures and to prevent static changes from affecting computers. Often space limitations and other important factors dictate the equipment to use; however, it is also important that the equipment not be undersized with respect to an adequate water treatment program. The following actual case illustrates the difficulties encountered when water treatment is relegated to an afterthought.

Boilers: Two, at 4500 lb/h of steam (Fig. 68); internal water capacity of approximately 50 gal; high-pressure, steam-fired

Water: Softened Lake Ontario water

At a 100 percent load, which was normal for the boiler, the unit evapo-

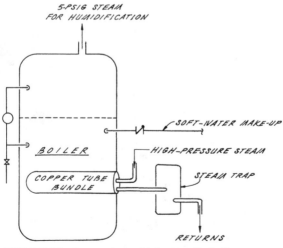

FIGURE 68 Low-volume, high-capacity boiler.

rated 540 gal/h of water, or one volume of boiler water every 5.6 min. The softened water used dictated a maximum of 15 cycles as solids, and alkalinity would be quite high by then. It can be seen that the maximum solids level would be reached every 83.3 min (1.4 h) of operation. This would require the operators to have the unit under constant surveillance or to maintain a constant surface blowdown of about 1.0 gal/min whenever it was in operation.

History showed the boiler waterside coils plugged solid with lime every year since installation. At times the water softener would malfunction and, because of insufficient workforce, not be noticed, allowing lime to build up on the tubes. Even though a boiler treatment compound was used, it could not keep up with the very high solids buildup in such a short time. Compounded with the problem of insufficient workforce, the boiler treatment compound was not effective in preventing lime buildup. The design engineer should have consulted with an experienced water treatment chemist, explaining the demands on the boiler, instead of relying on a water softener to solve all the problems. At the onset, one of the following suggestions could have been made:

1. Installation of an automatic, timer-actuated, conductance blowdown controller, as shown in Fig. 67.

2. Installation of a continuous surface blowdown controller, as shown in Fig. 69*A*, or a blowdown cooler, as in Fig. 69*B*.

3. In lieu of connecting to boiler control, installation of a 60-min repeating timer to the blowdown valve in Fig. 69*A*.

4. Since the job site was being fed high-pressure steam from a central steam station, a great deal of condensate was available. So that condensate could have been used as make-up water for the humidification boiler.

With items 1, 2, or 3 one would have had blowdown, and thus solids control, whenever the boiler was operational. The same initiating circuit, via a contractor, could also operate a chemical feed pump to maintain proper waterside parameters. With item 4, one would not have had scale problems and, with a water treatment program, minimal corrosion, if any.

Medium-Pressure Boilers

These units are normally process boilers and thus have varying amounts of make-up water as well as unique problems on the job site. Because of the nature of their operation, we cannot use the same water treatment formulas as are available for the low-pressure units; we must use a water treatment program designed specifically for the operation. The water treatment contractor must take all factors into consideration, such as

1. Pretreatment of the make-up water by a water softener, dealkalyzer, and deaerator

2. Discharge of boiler chemicals into the environment via blowdown

3. Process peculiarities, such as whether the steam is used for laundry, food processing, or process heat via coils

The water treatment program should be designed to minimize corrosion — less than 5 mils/yr would be a good target — prevent pitting and scale formation, and not unduly contribute to high TDS through

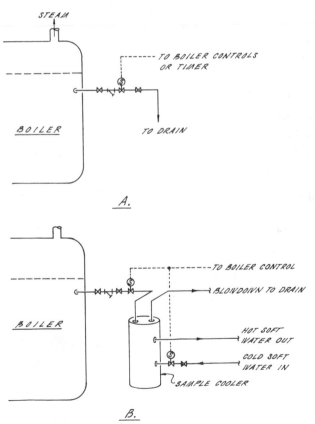

FIGURE 69 Continuous blowdown for boiler shown in Fig. 68.

overuse of treatment, such as sulfite. We would also like to minimize return-line corrosion but must exercise caution in the use of volatile amines.

Often a plant utilizing medium-pressure boilers will have incorporated into the specifications the desired water treatment program. You will often note the following parameters:

1. The pH shall be maintained within 10.5 to 11.5.

2. Alkalinity shall not exceed 700 ppm.

3. Phosphates shall be maintained within 30 to 60 ppm.

4. Sulfites shall be maintained within the 30 to 60 ppm.

5. Total dissolved solids shall not exceed X ppm.*

Yet the specifications will fail to provide the correct means to achieve the desired end. To meet the specifications, you must insist that the water treatment contractor supply two chemical feed pumps, two agitators, and two tanks per boiler as well as one set-up for feeding sulfite to the storage section of the deaerator, or return tank. It is almost impossible to maintain the specified parameters with a formulated single product or by adding pure products to a single tank. If the specifications do not require the three tanks and pumps, you can be sure that the competitive nature of the marketplace will ultimately find the plant with only one chemical feed pump and perhaps no tank if the chemicals are pumped directly from the drum. Figure 70 shows a recommended installation per boiler that will help the water treatment contractor meet the desired specifications. The blow-down controller is an optional piece of equipment if the boiler will not utilize continuous skimmer blowdown; however, it should not be used above 200 psig in a direct connection as shown. Above 200 psig you could utilize continuous blowdown with a heat-recovery heat exchanger to re-move heat from blowdown and use it elsewhere.

A problem with any chemical feed system is the chemical feed pump initiator. If the boilers are not under constant surveillance, how are you to know when the parameters are within bounds? Let us look at various initiators and note their pros and cons.

1. *On/off switches off boiler contactor to each chemical feed pump*

   This is a very good approach as long as you do not forget to turn it off or on, as the case may be. If it is left on for too long, the boiler will be

---

* The value is in thousands of ppm.

subjected to a great deal of treatment and thus TDS, which will cause operational difficulties such as priming, carryover, and constant low-water alarms. If left off, the consequences of an unprotected boiler are well known.

2. *Repeating timers off boiler control to contactor for each chemical feed pump*

   Like item 1, this is a very good approach as long as the timer is watched. The tendency here would be to have wide fluctuations in treatment parameters as the steam load varies, but the fluctuations would be acceptable.

3. *Reset timers off the boiler feed pump or flow switch to contactor for each chemical feed pump*

   This approach is very good for the phosphate and caustic but not for the sulfite feed. Sulfite use is dependent on the amount of oxygen entering the system, not on the amount of make-up water coming in. Further, in a multiple-boiler installation, which boiler feed pump would you use to activate the sulfite pump?

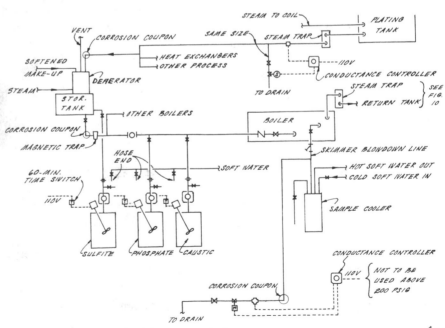

FIGURE 70 Chemical feed for process boiler.

4. *Flow switches to contactor for each pump*

The contact would be of too short a duration to allow sufficient treatment to be injected. This is not to say that flow switches cannot be used, but that they are not suitable as direct initiators.

The common approach is as described in item 1 — as the boilers come on, the chemical feed pumps are actuated. With this set-up one finds chemical testing being done at least once per shift.

A significant problem in treating boilers using sulfite for oxygen control is the control of the sulfite reserve in the boiler water. With one sulfite pump it is most difficult to maintain a correct sulfite reserve in a multiple-boiler installation unless the pump is manually adjusted each time a boiler is put on or taken off the line. The budget permitting, perhaps the installation of an additional, inexpensive sulfite feed pump per boiler, injecting into the storage tank, would be desirable. As each boiler comes on, its sulfite pump would also be activated.

Can a water meter installation such as that utilized for recirculating water systems be used (Fig. 55)? Unless the boiler operates on 100 percent make-up water, i.e., no returns, the use of a water meter is not feasible. Even with 100 percent make-up water, the water meter does not function well in controlling treatment addition.

As for the TDS, this parameter cannot be ascertained with the tools available to boiler room staff or, for that matter, to some water treatment companies. Presently TDS is ascertained as follows:

1. *TDS/conductance meters*

As discussed in Chap. 3, "Open Systems," pages 82–85, there are no instruments on the market capable of ascertaining dissolved solids. With boiler water you are also faced with the problem of alkalinity — the higher the alkalinity, the higher the indicated TDS or conductance. As compensation, you can add an organic acid, e.g., gallic acid or an aspirin tablet, to neutralize the alkalinity; but the reading thus obtained would be a guesstimate at best. .

2. *Hydrometers*

The suppliers of these instruments make the assumption that there is a direct relationship between total dissolved solids and specific gravity, with specific gravity defined as follows:

$$\text{Specific gravity} = \frac{\text{weight of substance}}{\text{weight of equal volume of water}}$$

The assumption that specific gravity and TDS are related is false, as attested to by the specific gravities of the following 10% solutions of pure products:

Product	Specific gravity @ 20°C
Soda ash (Na_2CO_3)	1.1029
Sodium chloride (NaCl)	1.0707
Sodium dichromate ($NaHCrO_4$)	1.0760
Stannous chloride ($SnCl_2$)	1.0810
Sodium potassium tartrate ($NaKC_4H_4O_6$)	1.0673

If there is no relationship with pure products, how can there be with a complex mixture such as boiler water? Figure 71 is a graph showing

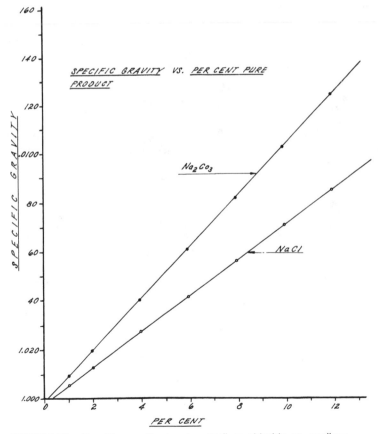

FIGURE 71 Specific gravity curve—sodium chloride vs. sodium carbonate.

specific gravity vs. percentage of two pure products — soda ash and sodium chloride. This should serve to prove that hydrometers are of no use as indicators of TDS.

3. *Cycles of concentration*

This method, while superior to those in items 1 and 2, is still unsatisfactory for determining TDS. This method can detect only those solids contributed by the make-up water, and then only by inference. For example, cycles of concentrations are found by dividing the chloride content of the make-up water into the chloride content of the system water:

$$\text{Cycles of concentration} = \frac{\text{chlorides in boiler water}}{\text{chlorides in make-up water}}$$

If you know the approximate TDS of the make-up water, you can multiply it by the cycles to find out how much TDS is being contributed by the make-up water supply. Note that this does not take into consideration the TDS contributed by the water treatment compounds, especially the sodium hydroxide and sulfites.

How, then, can you meet the specifications and stay below the specified TDS? If you are adamant about this parameter, you must insist that the TDS test be done by evaporation or an acceptable NACE or ASTM procedure that will provide a true TDS. A true TDS can be ascertained via the following procedure, done in triplicate:

1. *Filter a cooled sample of boiler water through a filter paper, Watman No. 42 or equivalent.*

2. *Fill a five-hundred- (500-) cm³ TC volumetric flask with the filtered sample.*

3. *Empty the filtered boiler water into a weighed six-hundred- (600-) cm³ flask. The flask shall be weighed to ±0.0005 gram.*

4. *Place the flask in a 190°F (87.8°C) ventilated oven, and evaporate the contents slowly over three to four (3 to 4) days.* ***Boiling is to be avoided!***

5. *On dryness, allow the flask to cool to room temperature in a desiccator and reweigh to ±0.0005 gram.*

   2(Weight of flask plus sediment − weight of flask)
   $$= TDS + \text{water of hydration}$$

6. *The flask may be placed in a vacuum chamber at 212°F (100°C) to remove the water of hydration and other volatile materials. This procedure will provide an accurate and real TDS figure.*

Is this amount of laboratory work and added expense really necessary? Minimizing the TDS reduces the possibility of priming, foaming, and carryover, so why specify an uppermost figure for TDS if it cannot be meaningfully ascertained in the field?

Rather than specify a particular program, perhaps you should consider the following specification for a medium-pressure unit:

WATER TREATMENT PROGRAM

The mechanical contractor shall obtain the services of a water treatment company that shall provide a water treatment program for a period of one (1) year from date of startup. The water treatment company shall be aware of any and all laws, rules, and regulations pertaining to the use of any treatment, chemical, or products supplied. Further, the water treatment company shall not supply any chemical, product, or products not in compliance with federal, state, or local laws, rules, and regulations pertaining to discharge of solid products into the environment or use contrary to OSHA rules and regulations. It shall be the responsibility of the water treatment company to ascertain compliance and agree to defend and hold blameless the design engineer, owner, and mechanical contractor against any action brought against them by the aforementioned authorities for noncompliance.

The water treatment company shall provide the services of a fully trained field engineer or representative, who shall provide the mechanical contractor with advice as to field installation of any and all equipment supplied by the water treatment company. The water treatment company's representative shall provide startup assistance as well as train the owners' maintenance staff in the proper application of chemicals as supplied. The representative shall have no less than an A.A.S. degree in chemistry, from an American Chemical Society (ACS) accredited institution, and five (5) years of experience in water treatment chemistry. Further, the representative shall be fully knowledgeable about problems related to under- and overuse of products as supplied.

The chemicals supplied shall perform in the following manner:

1. *General corrosion*

   As measured by corrosion coupons, supplied by the water treatment company, the general corrosion rate shall be kept under five (5) mils/yr in the boiler and return lines. In lieu of the corrosion coupons, the water treatment company may use any approved NACE or ASTM procedure.

2. *Pitting*

   Pitting, or accelerated pinpoint corrosion, of boiler tubes and return lines will not be tolerated. Should the water treatment company find factors conducive to pitting, they shall be brought to the attention of the design engineer via a written communication. The report shall state the factors and suggested corrective measures to be taken.

3. *Scale formation*

The formation of scale in the boiler, tubes, or shell, whether due to carbonate, iron, phosphate, silica, or any combination of the same, will not be tolerated. On inspection, at contract's end, the boiler(s) shall be free of scale or any encrusting matter. Should scale be found, it shall be cleaned by the water treatment company at its own expense. The owner may provide assistance, if only to provide some experience for the maintenance staff in boiler descaling, but is not under obligation to do so.

4. *Sludge*

This problem shall be prevented via the use of sludge modifiers and blowdown as required. Should sludge be present at inspection time, it shall be cleaned at the expense of the water treatment company.

5. *Dissolved solids*

This parameter shall be maintained at that level known by the water treatment company to not cause operational difficulties, such as, but not limited to, priming, foaming, heaving, and carryover.

It is understood that the owner's representative, on being trained in testing procedures, will test and maintain the boiler water and condensate within parameters set by the water treatment company. Further, the owner's representative shall maintain a daily log of all treatment readings and seek the assistance of the water treatment company's representative on anything not understood. Should problems occur and the owner's representative not understand how to correct same, the water treatment company's representative shall be called upon to assist. At no time should poor conditions be allowed to exceed seven (7) working days without corrective measures being taken. Should the owner's representative fail to follow instructions or maintain a proper log, then the water treatment company shall not be held liable for any problems or failures.

The water treatment company agrees to provide the following:

1. All chemicals, product, or products to protect the boiler(s) and return lines as outlined.

2. Feeding equipment specified but not piping, fittings, valves, electrical conduit, boxes, wall plugs, fuses, and other miscellaneous equipment shown on drawings. The water treatment company shall list, in an orderly manner, the equipment being supplied.

3. All necessary test kits and log pads for the owner's representative to test boiler water and condensate. The pH shall not be measured by using pH paper or a wide-range indicator.

4. Instruction of the owner's representative in proper testing procedures. At completion of the instruction period, the water treatment company shall submit a report stating who was trained and the parameters trained in.

5. *Technical advice for the mechanical contractor on the installation of feeding equipment. Errors in piping shall be deemed to have been approved by the water treatment company and shall be corrected, if required, at its own expense.*

6. *A visit to the job site no less than once per month, when the unit is operational, to ascertain compliance with the water treatment program. Noncompliance shall be brought to the attention of the owner's representative and the design engineer in writing. Failure to report noncompliance shall be taken to mean the conditions were satisfactory.*

Clearly this specification does not inhibit the choice of a water treatment program, nor is the design engineer called on to be a chemist. Performance is the goal; the specifics of the water treatment program are not the issue. With the push to limit the introduction of certain inorganic chemicals (phosphates and chromates, to name just two) into the waterways, water treatment companies are using more organic chemicals, not because they are safer or better but because the choices are limited only by a chemist's imagination. So why tie the chemist's hands by specifying a particular program? *Specify performance,* and let the chemists figure out how to achieve your ends.

As for chemical feed equipment for these process units, Fig. 70 is an ideal set-up, although it is a standard program of sulfite, phosphate, and caustic. The design engineer need not label or specify what the chemicals are for; just list them, and let the water treatment company decide whether the pump/tank/agitator set-up is satisfactory for the submitted water treatment program. However, the water treatment company should not be allowed to propose, in lieu of the standard multiformula program, a formulated product containing sulfite, phosphate, and soda ash with claims that the product is made specifically for the job. Certain allowances can be made in chemical feed equipment for a truly unique organic program, but not for a formulated product.

Should the design engineer care to be specific and list, not as an inflexible demand but as a guide, the standard water treatment program, then the following alteration of items 1 to 4 may prove helpful.

WATER TREATMENT PROGRAM

1. *A sufficient quantity of caustic soda shall be supplied to bring and maintain the pH of the boiler water to within 10.5 to 11.5 at all times.* * *Soda ash, Na_2CO_3, will not be permitted because it will break down in the boiler to*

---

* Near the upper limits of the medium-pressure line, 600 psig, it may not be desirable to maintain the pH at 10.5 to 11.5. It would be best to be guided by the boiler manufacturer.

generate CO_2, which will react with the condensate to form carbonic acid, which, in turn, will attack the return line.

2. *The total alkalinity in the boiler water shall not exceed* __[use limit or range dictated by boiler manufacturer]__ *ppm, expressed as calcium carbonate.**

3. *A sufficient quantity of a sulfite-based product shall be added to the system to keep the sulfite within thirty to sixty (30 to 60) ppm, expressed as Na_2SO_3, in the boiler water at all times.*

4. *A sufficient quantity of a phosphate-based product shall be added to each boiler to keep the phosphate within thirty to sixty (30 to 60) ppm, expressed as Na_3PO_4, in the boiler water at all times.*

5. *Sludge-modifying agents shall be added to maintain the sludge in a free-flowing and nonadherent state for ease of removal via blowdown.*

6. *Return-line corrosion shall be controlled via the addition of a mixed volatile amine product in such quantities as to maintain the pH of the return condensate within 7.5 to 8.0 at all times. The product shall not be added in a slug basis, nor shall the pH be allowed to exceed 8.0 at any time other than during malfunction of equipment.†*

7. *Dissolved solids shall be maintained at that level known by the water treatment company to not cause operational problems such as, but not limited to, priming, foaming, heaving, and carryover. Antifoam agents may be added as part of items 1, 3, 4, or 5 in order to minimize problems.*

These specifications shall serve as a basic guide along with the drawings. The water treatment company shall bid this job according to the plans and specifications; however, a better program may be proposed. The program shall be clearly marked "not per plans and specs," if that is the case. If the bid is accepted, the water treatment company shall issue a credit or refund to the owner for the value of any deleted item(s).

High-Pressure to Supercritical-Pressure Boilers

These plants are usually designed by teams of professional engineers who have had years of experience working with high-pressure boilers. Also, the plant's in-house chemist can provide valuable input since the chemistry involved at high pressures is not the same as that at lower pressures. The following parameters have to be carefully controlled:

---

* The normal limit is 700 ppm; however, around the 600-psig range you may not want to maintain the alkalinity that high.

† Note: Under no circumstances are volatile amines to be used where steam is used for humidification or food processing, including products that will come into contact with food.

Silica: Depending on the pressure involved, silica can become volatile and cause steamside silicate scale problems. This is most pronounced on turbine blades and similar equipment.

Dissolved solids: As pressures increase, the level of dissolved solids that can be tolerated drops significantly. This parameter must not exceed the limits set by the boiler manufacturer.

Suspended solids: Since the make-up water is carefully watched, you have to watch those solids contributed by the return water, e.g., copper and iron. The budget permitting, and as needed, you should consider the installation of magnetic traps on the return lines and polishing the condensate via ion-exchange resins.

Dissolved gases: As with other boilers, you must not allow the introduction of gases into the system. At the elevated temperatures involved, oxygen will make quick work of boiler tubes. As for carbon dioxide, the corrosive action will allow rust to come back to the boiler room via return lines.

Treatment The treatment program and the raw chemicals for these units are chosen so as to minimize the introduction of ingredients detrimental to the system. Obviously the program specified for medium-pressure units would not be suitable for high-pressure units since that would contribute to a dissolved solids problem.

Pretreatment is the key for high-pressure units. Depending on the source of the water supply, the use of a demineralizer is mandatory, with a reverse-osmosis unit a possibility in order to reduce loading on the demineralizer. Once the water is pretreated, it is fed to a deaerator along with the return lines, which hopefully have been polished. To go to these lengths and then introduce chemicals that will increase the TDS would be counterproductive, so you may find, for example, high-pressure plants using hydrazine as a primary water treatment chemical. Hydrazine will not contribute to the TDS during oxygen removal:

$$\text{Sodium sulfite} + \text{oxygen} = \text{sodium sulfate}$$
$$Na^+SO_3^{2-} \quad + \quad \tfrac{1}{2}O_2 \qquad Na^+SO_4^{2-}(s)$$

$$\text{Hydrazine} + \text{oxygen} = \text{water} + \text{nitrogen}$$
$$N_2H_4 \quad + \quad O_2 \qquad H_2O + N_2(g)$$

In this respect, hydrazine is slightly better to use than sodium sulfite; however, hydrazine does react significantly more slowly with oxygen.

There are catalysts on the market that will speed up the hydrazine-oxygen reaction, but it will still be slow. Above 550°F (287.8°C) (1030 psig) you cannot use sodium sulfite because it will decompose into undesirable elements — sulfur dioxide and sodium sulfide. Sulfur dioxide, on reaction with water (condensate), will form sulfuric acid.

The biggest problem in water treatment that a design engineer may face with a high-pressure plant lies not in the plant per se, but in the peripherals. So much attention is focused on the steam plant that the events at point use tend to be overlooked. For example, the author has had bids approved for the use of a nitrite/borate formulated product in a hot-water heating system served by steam from a high-pressure boiler. It did not occur to anyone that a converter malfunction could introduce contaminants into the condensate system. On learning about the nature of the system, the author suggested the use of hydrazine for corrosion control in the closed hot-water system in lieu of the formulated nitrite product. *Admittedly, hydrazine is not the chemical of choice for corrosion control in closed-loop systems. But you cannot afford to overlook the consequences of a converter malfunction on the entire system.*

A specific set of water treatment specifications applicable to high-pressure units is not realistically possible because each plant is unique in its operation. You could use the standard set of specifications up to about 600 psig; then alterations are necessary. But let us list items which would prove valuable to a water treatment company in the decision-making process:

1. Pretreatment measures taken: List everything that is to be done to the water as it enters the plant and prior to its use as boiler feed.

2. Maximum parameters listed by boiler manufacturer.

3. Use of steam:
   a. Power generation — how? Through what type of equipment?
   b. Process — define the nature of the process! You, the design engineer, know what use you are putting the boiler to, so why have the water treatment company guess?
   c. Heating — if through an HX, have you considered the consequences of an HX malfunction? Should a tube fail, you have to expect the closed-loop water to enter the return-line system. How will this affect boiler operation?
   d. Heating — domestic water.

Should you desire to generate specifications for a high-pressure plant, the following format may prove useful. However, these specifications will not be appropriate for critical-pressure units.

LI - 500 BH P
Druy - 4-9.30-2

WATER TREATMENT FOR HIGH-PRESSURE PLANTS

The mechanical contractor shall obtain the services of a water treatment company that shall provide a water treatment program for a period of one (1) year from date of startup. The water treatment company shall be aware of any and all laws, rules, and regulations pertaining to the use of any treatment, chemical, or products supplied. Further, the water treatment company shall not supply any chemical, product, or products not in compliance with federal, state, or local laws, rules, and regulations pertaining to the discharge of said products into the environment or use contrary to OSHA rules and regulations. It shall be the responsibility of the water treatment company to ascertain compliance and agree to defend and hold blameless the design engineer, owner, and mechanical contractor against any action brought against them by the aforementioned authorities for noncompliance.

The water treatment company shall provide the services of a fully trained field engineer or representative, who shall provide the mechanical contractor with advice as to field installation of any and all equipment supplied by the water treatment company. The representative shall provide startup assistance as well as train the owner's maintenance staff in the proper application of chemicals as supplied. The representative shall have no less than an A.A.S. degree in chemistry, from an ACS-acredited institution, and five (5) years' experience in water treatment chemistry. Further, the representative shall have no less than one (1) year's experience in treating units of similar size and pressure.

The make-up water to the deaerator shall be per the following analysis:

  [List here the analytical results as well as the name and address of the laboratory supplying same.]

The steam shall be used for the following purposes:

  [List here the use to which the steam will be put — be specific: humidification, through coils, HXs, whatever.]

And it is estimated the condensate return loss will be __[list]__ percent during __[list]__ process. The water treatment company shall be guided by the following recommendations from the boiler manufacturer:

  [List here all parameters as provided by manufacturer.]

The water treatment company is free to deviate from the recommendations if a better program is available; however, this will not absolve the water treatment company from providing the program recommended by the boiler manufacturer. Any program submitted shall provide the following protection:

1. *General corrosion and pitting*

   Pitting will not be tolerated, and general corrosion shall not exceed __[list]__ mils/yr as determined by NACE or ASTM procedure.*

---

* Make certain you, the design engineer, include the means for ascertaining the corrosion rate in the drawings. It makes little sense to specify a performance and then not provide the means for measuring it. Have someone in your office contact NACE or ASTM for the latest procedure in measuring the corrosion rate in the area you wish to monitor. Although corrosion coupons may be useful for return-line systems and low-pressure boilers, their use in medium- and high-pressure units may not be suitable.

2. *Suspended solids*

   This parameter shall be kept below the level recommended by the boiler manufacturer, __[list]__ ppm, as determined by __[list]__ procedure.

3. *Total dissolved solids*

   The TDS levels shall not be allowed to exceed the maximum limit set by the boiler manufacturer. Since TDS is an important parameter, it shall be determined by an NACE or ASTM procedure or by using the weighing method. Conductance shall not be used to estimate the TDS unless approved as such by owners.

It is understood that the owner's representative, on being trained in the testing procedures, will test and maintain the boiler water and condensate within specified parameters. Further, the owner's representative shall maintain a daily log of all hourly readings on treatment reserves in the water. Should conditions not be satisfactory, the owner's representative shall contact the water treatment company's representative for advice on means to correct conditions. At no time shall poor conditions be allowed to continue for longer than forty-eight (48) hours. Should the owner's representative fail to follow instructions or maintain the proper log records, the water treatment company shall not be held liable for any problems or failures.

The water treatment company agrees to provide the following:

1. All chemicals, product, or products to protect the systems outlined as specified.

2. Feeding equipment specified. The mechanical contractor shall be responsible for its installation and for supplying the piping, valves, fittings, and other miscellaneous items shown on drawings. The mechanical contractor shall also be responsible for the electrical wiring shown.

3. All necessary test kits and log pads needed for the owner's representative to test boiler water and condensate. The pH shall be ascertained by using a laboratory-grade pH meter — Corning, Beckman, Fisher Scientific, or approved equivalent.

4. Instruction of the owner's representative in proper testing procedures. At completion of the instruction period, the water treatment company shall submit a report stating who was trained and the parameters trained in.

5. Technical advice for the mechanical contractor on the installation of feeding equipment. Errors in piping shall be deemed to have been approved by the water treatment company and shall be corrected, if required, at its own expense.

6. A visit to the job site no less than __[list]__ per month when the unit is operational to ascertain compliance with water treatment program. Noncompliance shall be brought to the attention of the owner's representative and the design engineer in writing. Failure to report noncompliance shall be taken to mean that conditions were satisfactory.

Note that the specifications writer does not accept the onus of responsibility for the treatment program. The boiler manufacturer's metallurgists and chemists are in a better position to provide the necessary information since they are acquainted with the peculiarities of the steel being used. They, together with the chemists from the water treatment company, should be able to come up with a specific set of parameters to prolong the life of the boiler at a reasonable cost.

Q. what's A closed syst?
is it ""OPEN syst?

1 GGC

spring cleve

neural classification

669

SIX

Domestic Water Treatment

In this chapter we are concerned with the water as it enters the plant. This supply, referred to as *raw,* may or may not be suitable for in-plant or process use. The water, as obtained from its natural source, need only be filtered and chlorinated to meet the potable needs of the community; rarely will it meet all plant needs. If a plant is located in an area because of its water supply, e.g., hard or soft, provisions will have to be made to alter the chemical characteristics of the water to meet the plant's domestic and HVAC needs. Let us address these needs and use the following definitions:

Soft water: No scale formation on the tubes of a domestic hot-water heater is noted when water is heated to 190°F (87.8°C).

Hard water: Scale formation on the tubes of a domestic hot-water heater is noted when water is heated to 120°F (48.9°C) or higher.

Very hard water: The cold-water supply turns milky if allowed to warm to room temperature.

In the past it has been customary to refer to these supplies with specific figures. However, the author will deviate and base his definitions on the performance of the water at specific temperatures. There are too many gray areas with figures which are subject to interpretation, but specific results are clear-cut and independent of other variables. For example, a water supply of 100-ppm hardness could be considered hard in some areas and yet soft in others, as a result of the alkalinity difference in the supplies. In the traditional sense of the word, *hardness* refers to the ability of the water to form curds with soap. If it does, it is hard; if it does not, it is soft.

SOFT SUPPLY

Figure 6, page 15, case 47348, is an example of a soft-water supply even though it contains 42-ppm total hardness. Even at 190°F (87.8°), where $I_s = -0.72$, this supply will not form lime on the tubes of heat exchangers. At cold-water temperatures, about 40°F (4.4°C), $I_s = -2.16$ and as such is very corrosive. It is obvious that this supply must be treated, by the supplier (if the life of the distribution system is to be maximized) or by the point source user.

Since we are dealing with a supply with, perhaps, five end-uses — process, steam boiler, air conditioning, domestic hot water, and domestic cold water — we must adjust the chemical characteristics to suit each need. Let us assume the supply is perfect as is, at 40°F (4.4°C) for process use. The parameters to be maintained for boilers and HVAC systems were covered in other chapters. We are left, then, with the necessary corrections for the domestic water system.

The domestic water system, with its two temperature extremes, does pose somewhat of a dilemma. Let us assume the corrective chemical to use is calcium hydroxide, $Ca(OH)_2$, in order to bring I_s as close to zero as possible. To what I_s do we adjust? If we add sufficient chemical to protect the 40°F (4.4°C) system, we will make I_s too positive at 180°F (82.2°C), which will allow lime to build up on HX tubes. If we correct for the 180°F (82.2°C) system (which will not require much in the way of adjustment since the saturation index is only -0.78), the 40°F (4.4°C) system will continue to corrode and leach metals from the system. The solution to this perplexing problem is to treat these two as separate supplies, as shown in Fig. 72.

The two electronic chemical feed pumps are controlled by pulses from the water meters as the volume varies — this is a must. We are trying to maintain close control over those parameters which affect I_s. Figure 72 shows a single chemical feed tank with one agitator; however, two may be required. As is evident, this is a very simple system which is easy to maintain, assuming the chemical used is not prone to bacterial attack or reversion, such as is possible with polyphosphates.

The following specifications should assist the engineer in setting up a program to extend plumbing and equipment life.

DOMESTIC HOT- AND COLD-WATER SYSTEMS

The mechanical contractor shall obtain the services of a water treatment company that shall provide a water treatment program for a period of one (1) year from the date of startup for the domestic water systems. The water treatment company shall be aware of, and be responsible for, any and all health rules and regulations pertaining to the use of any treatment, chemical, or products supplied. The water treatment company shall not supply any chemical, product, or products not in compliance with federal, state, or local laws, rules, and regulations pertaining to the ingestion of the products to be supplied. Further, the water treatment company agrees to defend and hold blameless the design engineer, owner, and mechanical contractor against any action brought against them for the use of the recommended product or products. It is understood that the owner's representative, on being properly trained in the test procedures, will test the domestic water system and maintain the chemical characteristics within parameters determined by the water treatment company.

The chemical treatment tank and lines shall be cleaned in the manner and frequency prescribed by the water treatment company in order to maintain the tank, pump, and lines free of all living unicellular and multicellular organisms.

The water treatment company shall provide the services of a fully trained engineer or representative, who shall offer the mechanical contractor advice as to field installation of any and all equipment supplied by the water treatment company. The representative shall provide startup assistance as well as train the owner's maintenance staff in the proper application of chemicals as supplied. The representative shall have no less than an A.A.S. degree in chemistry, from an ACS-accredited institution, and five (5) years' experience in water treatment chemistry.

The water treatment company shall provide a food-grade treatment chemical to protect the domestic water systems listed against scale and corrosion. Corrosion is to be maintained at less than five (5) mils/yr as determined by corrosion coupons or approved NACE of ASTM procedure.

The chemical make-up of the raw water, as supplied by ___[list source]___ , is as

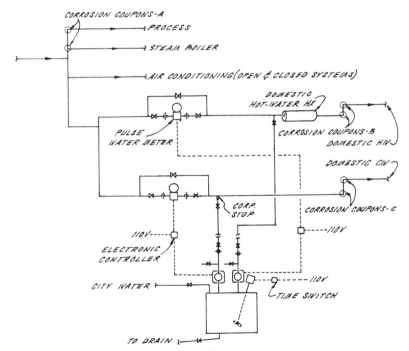

FIGURE 72 Domestic water chemical feed system.

follows:

pH   _____

P alkalinity, ppm as $CaCO_3$ _____

M alkalinity, ppm as $CaCO_3$ _____

Total hardness, ppm as $CaCO_3$ _____

Chlorides, ppm as NaCl   _____

Conductance, micromhos   _____

TDS by meter, ppm   _____

TDS by weight, ppm   _____

The water treatment company is free to conduct its own tests, and should analytical data differ from those supplied, it shall be so noted. However, the water treatment company is obligated to calculate and recommend based on the information supplied.

The amount of treatment supplied shall be based on the following flow rates:

Cold water   _____ gal/min or _____ gal/day

Hot water   _____ gal/min or _____ gal/day

The equipment shown in the accompanying drawings shall be supplied by the water treatment company only and shall consist of:

1. *Two pulse-type water meters to activate chemical feed pumps in direct proportion to water flow.*

2. *Two electronic chemical feed pumps of adequate capacity, interconnected with the water meters, to inject a proper quantity of treatment into each system. The pumps shall be piped as shown in the accompanying drawings.*

3. *Two corporation stops.*

4. *One chemical treatment tank of ___[list]___ gallon capacity.**

5. *An agitator and timer, as shown in the accompanying drawings, if a dry treatment is offered.*

---

* The greater the water consumption, the larger the tank has to be, especially if you are using only one tank. For health reasons it would not be wise to specify a tank with a capacity such that at peak load the contents will last longer than two (2) weeks.

6. *Six corrosion coupon holders* and sixty (60), weighed to ±0.0002 gram, soft-steel corrosion coupons as shown in the accompanying drawings.†*

7. *The necessary test kits to ascertain:*

   a. Langelier's saturation index

   b. Amount of corrosion inhibitor in system

   Test kits shall be Hach, Hellige, Taylor, or equivalent, but shall be of same manufacture.

8. *Sufficient log pads and sample containers for mailing water samples to water treatment company's laboratory.*

9. *Laboratory analyses and evaluation of test coupons.*

10. *Onsite visits of the representative no less than once per month during contract period to ascertain conditions and evaluate logs. Deviations from proper conditions shall be brought to the attention of the owner and design engineer in writing.*

As with other specifications, *we are requesting performance* rather than providing set objectives, i.e., chemical parameters. We could specify a certain I_s range, say, ±0.2, but that would not guarantee that we would achieve the desired end — minimum corrosion. It would be wise to let the water treatment company decide what is best for that particular water supply. Once the objective is stated, it is up to the water treatment company to achieve it, especially when adding chemicals to potable water systems.

HARD SUPPLY

Let us now consider a hard supply per our definition — a scale former at 120°F (48.9°C). Figure 6, case 16300, fits our description very well. With a total hardness of 705 ppm (41.3 gal/day), some engineers would classify this supply as atrocious, awful, or very high, all meaningless for our purposes.

This supply is not too bad at 40°F (4.4°C), where $I_s = +0.5$, but it starts to become a problem at 120°F (48.9°C), where $I_s = +1.42$ and is impossible at 180°F (82.2°C), with an I_s of 1.88! At the cold end, we should not expect too many problems; but at 180°F (82.2°C) the water must be softened! The

---

* To be installed in system at locations *A*, *B*, and *C* as shown in Fig. 72. The actual location is not too important as long as you have a corrosion rate for untreated and treated water to measure program effectiveness.

† In this book, see Fig. 28 for dimensions.

common approach has been, and continues to be, to soften the make-up water to the domestic hot water system 100 percent. *Is this approach good? It is acceptable, and it is as old as water softening. But there is a better way — partial softening!*

With partial softening we mix soft water having an I_s of -0.84 with raw water having an I_s of $+1.88$, to come up with an I_s close to zero or just very slightly negative at 180°F (82.2°C). The final mix could have the following approximate chemical characteristics:

pH	7.7
Total alkalinity, ppm as $CaCO_3$	225
Total hardness, ppm as $CaCO_3$	8.0
TDS via meter, ppm	607
I_s	-0.02

Figure 73 shows a recommended installation for softening and mixing.

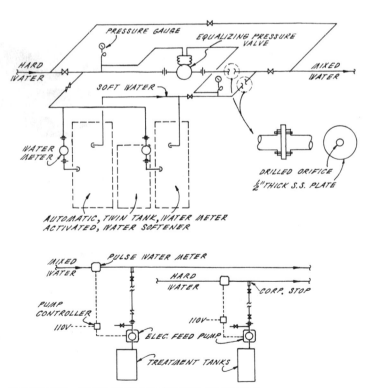

FIGURE 73 Partial mixing and treatment addition.

The drilled orifice, whose diameter is determined by calculating the final water mixture desired, is installed as shown. The author knows of no reason why ball valves cannot be used in lieu of the orifices, other than the reduced flow rate resulting from increased resistance. The equalizing pressure valve is used to maintain constant pressure to allow for constant flow and thus maintain an almost constant mix after the orifices.

The advantages of this approach are as follows:

1. The final mixed supply should not be as corrosive to the plumbing as the 100 percent soft water supply.

2. Unsightly blue-green (copper) and brown (rust) stains in sinks will be minimized.

3. Laundry staining caused by rust will be controlled.

The disadvantages are the extra equipment required and the chemical testing that must be performed to maintain the water within design parameters.

Is this approach cost-effective? It is a difficult question to answer without addressing the impact that a water supply with an I_s of -0.84 has on the plumbing. What is the corrosion rate of water with an I_s of -0.84? Actually we want to ascertain the life expectancy of the system plumbing vs. that of one with an I_s of -0.02. Let us not forget to interject into our equation the labor hours that will be spent repairing corroded system components as well as the cost of replacement parts.

As insurance against corrosion, or scale formation, we should add treatment to the finished supply. The chemical feed system would be as outlined in Fig. 73, with a water meter activating an electronic pump based on the volume of water. The same specifications as outlined on pages 154–157 for soft water can be used. The only addition would be the size of the water softener and drilled orifice or the use of ball valves.

VERY HARD SUPPLY

A very-hard-water supply, as we defined it, has to be softened for domestic water use — there is no choice! However, what many people fail to realize is that on softening, a visible annoyance — scale on HXs — is changed to an invisible one — corrosion! We saw how the hard-water supply went from an I_s of $+1.88$ at $180°F$ ($82.2°C$) to an I_s of -0.84. The same will happen with very hard water. The author has a client with a building having a severe corrosion problem, all because it was decided at the design stage to soften all the water in the building. It has taken a few years, but corrosion is now taking a toll of the plumbing and piping system.

If complete water softening is not the answer, then perhaps, as with the hard water supply, we could utilize partial softening for the hot water and chemical control for the cold supply. Partial softening has already been discussed and its value established. Thus let us turn our attention to chemical control for the cold supply.

By definition, very hard water is saturated with respect to lime, which explains why the water will turn milky on standing. The only chemical approach to the problem is to prevent the lime from coming out of solution. This can be accomplished via the addition of the following:

1. Sequestering agents to hold the hardness in solution. A common agent would be polyphosphates at a dosage, around 3 to 5 ppm, appropriate to accomplish the task. It is important in using polyphosphates that the water be tested for that ingredient, and not for the *ortho* form. It is also important not to overfeed, since that can lead to formation of a phosphate sludge in the system.

2. A food acid to alter the LSI by the reduction of the alkalinity. It is possible that the alkalinity-hardness ratio is such that by lowering the alkalinity the precipitation of lime will cease. We are not going to make the water acid, just alter the alkalinity slightly. One should not start adding acid, even food-grade phosphoric acid, to a domestic water system for scale control without having a good survey of the

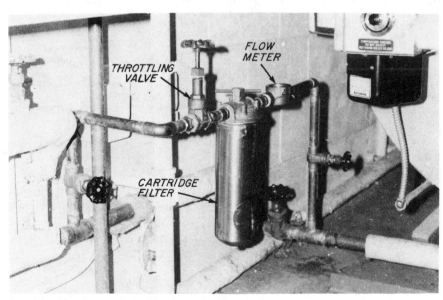

Cartridge filter in a recirculating domestic hot-water system.

Domestic water treatment using a by-pass feeder.

situation or an analysis to go by. This is only an option that can be considered under the right conditions.

As with any system under treatment, the results of the treatment program should be monitored. No piece of equipment is so perfect that it is failureproof. The following suggested log should be maintained on a weekly basis and the results compared with desired parameters:

| | | *Values* | |
| | | *Treated* | |
Parameter	*Raw*	*40°F (4.4°C)*	*180°F (82.2°C)*
pH	——	——	——
Total alkalinity, ppm as CaCO$_3$	——	——	——
Total hardness, ppm as CaCO$_3$	——	——	——
Conductance, micromhos	——	——	——
Chlorides, ppm as NaCl	——	——	——
LSI	——	——	——

For long-term monitoring, it would be advantageous to install corrosion coupons in each system. We must know what the corrosion rate is for the raw water to ascertain whether the treatment program is working. If the raw water has a positive I_s, and thus is not corrosive with respect to that index, its corrosion rate would not be too meaningful; however, it would be good to know.

VARIABLE SUPPLY

All the supplies covered so far have been assumed to be unchanging; i.e., the chemical nature stays constant year-round. We have not addressed the supply that varies. Figure 4, the case of the city of Ithaca, NY, is just such a supply. In 1981 the hardness varied from 98 ppm in May to 140 ppm in June; in 1982 it was 90 ppm in April and 188 ppm in October. For 1983 it went from 70 ppm in May to 275 ppm in June, just 30 days later! Figure 74 graphically illustrates the problem of where to start.

If we suggest a water softener for the entire supply, on what do we base our calculations for softener capacity? There is no doubt that we must use the highest figure obtained thus far, 188 ppm (11.0 gr/gal). If we estimate a daily water consumption of 25,000 gal, we must specify a water softener capable of softening 300,000 gr/day, which will prove to be an expensive item. Furthermore, will softening solve the problem? Let us see what the saturation index is for the worst possible supply (see Fig. 2):

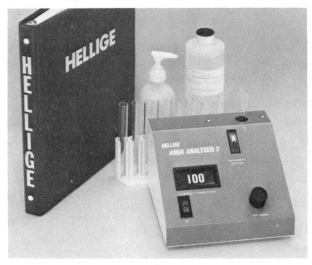

Hellige aqua analyzer. (Courtesy Hach Company.)

Parameter	Saturation index		
pH = 7.8	pCa	2.70	2.70
Hardness, 200 ppm	pAlk	2.55	2.55
Alkalinity, 140 ppm	Temp.	40°F (4.4°C)	180°F (82.2°C)
TDS, 200 ppm	C @	2.61	1.24
	pH$_s$	7.86	6.49
$I_s = $ pH $-$ pH$_s$	I_s	-0.06	1.31

We see that we have a slightly corrosive, non-scale-forming supply at 40°F (4.4°C), but scale will form at 180°F (82.2°C) because $I_s = +1.31$. Now let us see what happens to I_s as the water changes to a good supply:

Parameter	Saturation index		
pH = 7.1	pCa	3.15	3.15
Hardness, 70 ppm	pAlk	2.95	2.95
Alkalinity, 55 ppm	Temp.	40°F (4.4°C)	180°F (82.2°C)
TDS, 125 ppm	C @	2.95	1.24
	pH$_s$	9.05	7.34
	I_s	-1.95	-0.24

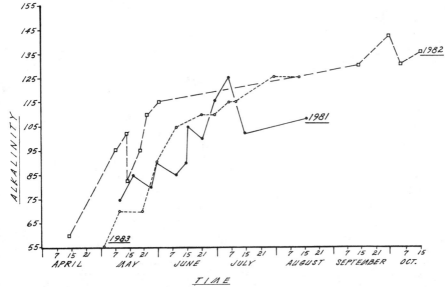

FIGURE 74 Alkalinity graph of chart shown in Fig. 4.

As we can see, we have gone from a slightly corrosive supply to a very corrosive one at 40°F (4.4°C), from an I_s of -0.06 to one of -1.95. The hot-water supply at 180°F (82.2°C) improved favorably. Apparently the users of this supply must live with extremes in their water that affect both the cold- and hot-water systems. This is not to say that nothing can be done. Let us examine alternatives.

Hot-Water Supply

Since a water softener is not really an option, we have to think in terms of chemical alteration. We cannot alter the hardness, but we can alter the alkalinity to change I_s. If we add an acid, as previously suggested and as shown in Fig. 75, we can change I_s and try to keep it in a favorable state. For example, let us add a food-grade acid and alter the worst possible supply as follows:

Parameter		Saturation index
pH = 7.1	pCa	2.70
Hardness, 200 ppm	pAlk	2.95
Alkalinity, 55 ppm	C	1.24 @ 180°F (82.2°C)
TDS, 200 ppm	pH_s	6.89
	I_s	0.21

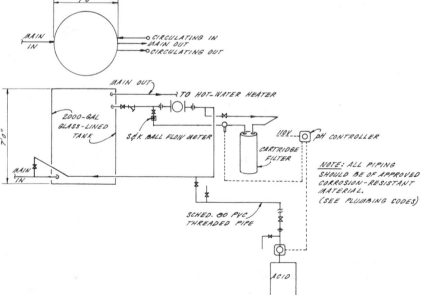

FIGURE 75 Alkalinity reduction for recirculated domestic hot make-up water.

As can be seen, we have not altered the hardness in any way, but have tamed the water by bringing I_s down from $+1.31$ to $+0.21$ just by adding a little acid. This is not to imply that this will be an easy procedure; it is just an option that can be utilized. Keep in mind that by altering the water for $180°F$ ($82.2°C$) use, we are making it very corrosive at $40°F$ ($4.4°C$), which is why in Fig. 75 corrosion-resistant material for the tank and all plumbing is required. Further, the water for the application must come as close as possible from the main if we are to treat the corrosive city water.

Cold-Water Supply

Since the cold-water supply has a negative index -0.06, at its worst and the index just gets more negative, -1.95, as the supply improves, we are faced with the problem of corrosion control, as we were with the soft water supply. However, here the problem is not so simple. The water supply's I_s is much too variable for direct injection of an I_s improver such as calcium hydroxide. It is possible that we could utilize the set-up shown in Fig. 75 but in lieu of a pH controller substitute a corrosion monitor. Acid would not be used; instead, we would inject a food-grade corrosion inhibitor on a continuous basis, altering the feed rate depending on the corrosion reading.

The recommendations in this section are only *possible alternatives* that could be investigated as a solution to the problems contributed by a variable water supply. The answers are not simple or clear-cut. You must analyze the supply, define the problems, and propose the means for correcting same. You must balance the cost of the proposed corrective measures against the cost of not doing anything.

WATER SOFTENING

The greatest problem associated with softening is not the water per se, but the water softener. When water softeners are being sold, the phrase "maintenance-free" seems to be stressed. However, nothing could be further from the truth! Water softeners require maintenance, both obvious, such as valve repairs, and not so obvious, such as attention to the resin.

When the water softener allows hard water to pass, the operator assumes more brine time is needed and thus increases the brining cycle. If the operator is performing daily chloride tests, which is not normally done or recommended by the sellers of the units, brine valve problems will be noticed before they become severe. If the operator is not performing the daily chloride tests, greater and greater amounts of salt will be injected along with the soft water. It is bad enough to convert a scale-forming supply to a corrosive one. Why compound the problem by allowing salt to enter?

With respect to valve bodies, some are constructed of steel and are very prone to corrosion. Some have been seen to crumble at the touch, especially at valve seat areas. Others are constructed of bronze, with steel machine screws that break off at the base when one tries to remove them, and plastic, with valve seats of Teflon. But there *is* one type, constructed of a thermosetting plastic, cam-operated, with seats and seals of Teflon, that is corrosion-resistant.

When a valve body is rebuilt and the softener continues to allow hard water to pass, suspicion is cast on the resin. Indeed, by the process of elimination, the problem *has* to be the resin. Here is the difficulty: How can you determine whether the lack of softening capacity is due to loss of resin or to reduction of resin surface area because of fouling? On cleaning water softeners, you will usually find the unit loaded with a white/gray sediment and perhaps some flocculent material. Often this old resin will be discarded and replaced with new — which, costing a little over $100 per cubic foot may cause you to wonder if there is still life in the old resin. And the resin frequently can be brought back by washing it thoroughly and then proceeding with heavy brining. The trade-off is worker-hours vs. cost of new resin. Should you want to try to revive some depleted resin, the following procedure can be used for each 3 ft³ of resin:

1. *Construct a direct brining set-up as shown in Fig. 76.*

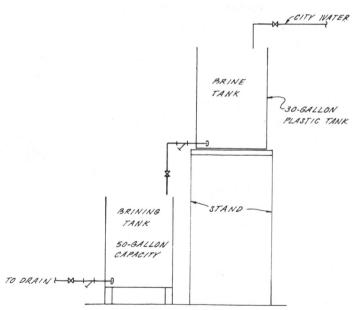

FIGURE 76   Direct brining of exhausted water-softening resin.

2. *Place clean, exhausted resin in the brining tank.*

3. *Add brine, a saturated solution of clean salt, until it reaches a level three (3) inches above resin.*

4. *With a paddle, or agitator, stir the mixture for a few minutes; then let stand for no less than one (1) hour.*

5. *Drain the brining tank, and test the drain water for hardness. When the hardness is equal to or less than that of city water, the resin has been cleaned as much as direct brining will allow. Continue procedures 4 and 5 until conditions are met.*

6. *Drain the brining tank, and add a sufficient quantity of a resin-cleaning product specifically formulated for the purpose. Let stand 24 hours, drain, and repeat the brining procedure once again. If the hardness of the drain water is equal to or less than that of city water, the resin is as good as new.*

Invariably, when the sediment is noted on cleaning the resin, operators want to know why softener back-flushing does not remove the sediment from the resin. *They do not realize that back-flushing is used only to remove the excess brine from the unit, not to remove physical sediments.* The sediment must be removed manually on a periodic basis. How periodic? This must be ascertained by the operator from the tests being performed.

Hach water hardness test kit, model 5B. (Courtesy Hach Company.)

Figure 77 gives an idea of when a water softener should be manually cleaned. If we accept a 5 percent nominal physical resin loss per year, it is easy to see that the softener will lose about 40 percent capacity in 10 years. If we then add a 2 percent loss due to fouling, the total loss will amount to over 50 percent, but that is over 10 years. If we study the graph and accept the 7 percent loss curve as reasonable, we will note that the water softener performance drops after the second year. If we have a 90,000-gr unit and regenerate at 80,000 gr continuously, by the third year the softener will be passing hard water as it approaches its regeneration time. This can be a problem if we are relying on 100 percent soft water, e.g., when we mix soft

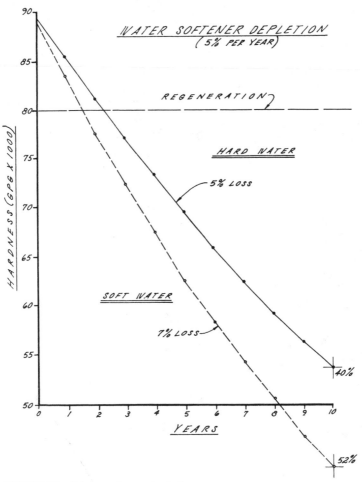

FIGURE 77 Water softener depletion curve.

and hard water to achieve a desired saturation index, for a particular purpose.

Now that we are aware of some of the field problems, how can we help future clients? First, we must take a critical look at our favorite water softener:

1. Is the valve body made of corrosion-resistant material?

2. Are operating valve parts easily serviced?

3. Is the brine tank corrosion-resistant? Note that iron fouling shortens resin life considerably, perhaps greater than the 2 percent shown in Fig. 77. An epoxy-coated steel tank for brine storage is really not the best approach, especially when excellent plastic ones are available.

4. Is the resin tank internally corrosion-resistant?

5. Can the resin tank be cleaned easily, with minimal equipment? The author and two men have struggled for 3 h to physically clean a 150,000-gal grain capacity resin tank. The danger of the tank falling on one of us was very real and could have resulted in severe injury. With the second tank the men wanted to know whether we could just flush the resin with a water hose rather than remove it. This should illustrate why ease of cleaning is a must.

As for a testing program, it is imperative to set one up and maintain it on both the hard- and soft-water supply. How else can we know whether a malfunction is imminent? The water for testing should be obtained upstream and downstream of the softener, not from the valve body, and should be tested for the same parameters as previously listed (see page 161).

It is hoped that this section alerts those that purchase and recommend water softeners to the potential problems faced in the field. Water softeners are vital, but they do require attention and service.

SEVEN

Chemical Safety

Chemical safety must be discussed since, with the advent of Rachel Carson's *Silent Spring,* chemicals have been on everyone's lips, physically and figuratively speaking.

How safe is that product? This question is asked with greater and greater frequency. It is multifaceted, to say the least, and presents problems for the manufacturer, formulator, seller, and user. The manufacturer of the pure raw materials — for example, the disodium phosphate found in some foods — is under obligation to test for product effectiveness, safety in handling, health precautions, and environmental impact. The manufacturer's data sheets are usually quite extensive and provide ample information for the formulator (the water treatment company). The formulator is concerned with safety during in-plant formulation. OSHA is trying to regulate, and unions are taking a very close look at, health hazards associated with the workplace and the listing of precautions per government regulations. The seller, the person who convinces you that you need this product, has the same basic concerns as the buyer, but the buyer's primary concern is safe handling. The end-users, the maintenance personnel, are concerned with immediate safety in terms of handling, spills, but also with environmental impact (can I, or can I not, drain the cooling tower bleed down the sewer?).

On the surface, most products seem safe and should not pose problems, but chemistry is not that simple. There is a world of difference between a safe, single chemical entity, e.g., sodium phosphate, and the chemical "soup" to which this is added to create a marketable product. One cannot, and should not, assume that these final products are thoroughly laboratory-tested for safety and environmental impact. More often than not, the end-user is the first to notice odd effects; then the formulator alerts field representatives to possible hazards. This is not to say that water treatment or other proprietary products (e.g., soaps, disinfectants, cleaners) are dangerous, but that you should *not* assume that a product is 100 percent safe. Keep in mind that economic considerations preclude testing the finished formulations, and thus the full impact on the environment and handlers cannot be appreciated. Drugs are supposed to be thoroughly tested, but even so bad effects from their use may take years to be noticed. For example, thalidomide, a widely used drug, was found, when used by women during pregnancy, to cause infant deformities. *Unlike drugs, the products used by maintenance personnel are not required to be tested by any government agency.* The rule is *caveat emptor,* or "let the buyer beware."

Let us take a superficial look at some ingredients[15] used in water treatment formulations and note the potential for problems.

AIR CONDITIONING SYSTEMS

Chromium

As a metal, chromium is found in low concentrations in some foods and is thought to be an essential diet element. Consequently, it is found in some vitamin and mineral supplements. But as a chromate, its oxidized state and the state in which it is used in water treatment formulations, chromium is thought to be a carcinogen. However, the *World Health Organization's International Agency for Research on Cancer (IARC) monographs only suggest that lung cancer is a risk* among workers in the chromate-producing industry, not unlike the risk faced by workers in the asbestos industry. *The monographs present no evidence that nonoccupational exposure to chromium increases the risk of cancer.* In laboratory tests on animals, rats, mice, rabbits, and guinea pigs, oral and inhalation administration of *chromates failed to produce cancerous growths.*

This is not to imply that chromates are 100 percent safe. The author has had firsthand experience with a severe case of dermatitis that allegedly occurred every time a particular maintenance man managed to get chromate treatment — the mixture in the treatment tank which is about a 10% solution — onto his skin. The dermatitis was real. However, was it caused by the chromate, a contaminant in the raw chemicals, or something which grew in the treatment tank? Since the tanks (i.e., chemical mixing tanks) are rarely cleaned and since polyphosphates were part of the treatment program, possibly something other than the chromates was responsible for the dermatitis.

It is interesting that *some products with a high carcinogenic tendency,* e.g., tobacco and nitrites in cold cuts and other foods, *are not illegal; yet, despite the lack of condemning evidence, chromates are prohibited.* Some communities demand that chromates be removed even if they have been in the water system for many years. The question still remains: *Are chromates dangerous?* The author believes *they are not* and still recommends their use for steam-heating boilers, closed hot- and chilled-water systems, and open recirculating water systems. Since federal limits on chromates are quite low where the waste, bleed, or blowdown enters waterways, care must be taken not to use excessive amounts of them. But in large municipalities where the waste enters the sewage system, the amount of chromate in the system due to bleed or blowdown is diluted so much that the federal limits are not even approached.

Nitrites

Nitrites are presently used as a preservative in cold cuts and in curing some meats. *It is believed that this compound, on ingestion, is involved in the formation of carcinogenic n-nitroso compounds.* However, available data have not proved nitrites to pose a significant health hazard to warrant legislation barring their use in foods. Since nitrites may be used in foods, their use by the water treatment companies continues. *Is this product safe?* Let us analyze some conditions and draw our own conclusions:

1. *Cooling tower use*

   Although nitrites are an excellent corrosion inhibitor, their *use* as such in open recirculating water systems is *limited because of their propensity for bacterial reversion. Nitrites are food for certain classes of bacteria.* When the reversion process starts, it is impossible to stop; you can use great amounts of algaecides and chlorine, but to no avail. Given the threat of Legionnaire's Disease, it would not be wise to add a compound to the system that could help increase the general bacterial population.

2. *Closed loops*

   a. *Chilled systems, 40°F (4.4°C)*

   The same reasoning applies to these systems as to the open ones. Once bacterial reversion starts, there is no cure. All the algaecides and sterilizing procedures are just an exercise in futility and extra profit for the chemical supplier. The author has had systems leak, after reversion had proceeded for quite some time, and *the water had a very strong odor.*

   b. *Moderate systems, 100°F (37.8°C) maximum*

   Because the temperature, 100°F (37.8°C), is so close to the human body temperature of 98.6°F (37°C), *the author does not recommend the use of nitrites in these systems.* If nitrites can grow and prosper in this system at 100°F (37.8°C), they can do likewise in moist lungs. This is not to say that only nitrites are suspect, but all bio-degradable formulations for closed-loop systems should be suspect.

c. *Hot-water systems, 140°F (60°C)*

The author has had very good results with the use of nitrites in these systems, at 1000 to 1500 ppm. Reversion has never been experienced, and the nitrite held, except for times when leaks prevailed. When nitrites are used in these systems, there must be a back-flow prevention device so that the domestic water system is not contaminated with nitrite-treated water.

Sulfuric Acid

There is no doubt that this chemical, at a certain concentration, is dangerous, but how dangerous and at what concentration? Let us look at some common concentrations and note the precautions.

60° Baumé The 60° Baumé sulfuric acid is also known as oil of vitriole and concentrated sulfuric acid. Even though purchased as 60° sulfuric acid, *at this strength it is not a true acid.* Nevertheless, this concentration, if spilled on cotton clothing, will dehydrate the item, i.e., take out the moisture and molecular water and leave behind carbon. The carbon residue leads one to believe that this acid burns. When 60° Baumé sulfuric acid is splashed in the eyes, the same reaction takes place; however, since the eyes are always moist, the reaction is quite vigorous and heat is generated. Under these conditions people can lose their sight. When it is spilled on dry hands, there is no noticeable discomfort — just a slight warmth where the acid draws the water from the skin cells. But when it is spilled on moist hands, crotch, back, or armpits, then there are problems. The reaction between the perspiration and sulfuric acid is so intense that the skin is actually burned, not by the acid but by the heat of reaction.

Under the right conditions, 60° Baumé sulfuric acid could liberate sulfurous fumes which could trigger respiratory problems such as asthmatic attacks.

The author urges that the piping arrangement shown for the use of sulfuric acid in open recirculating systems in this book be followed with no deviations. The piping arrangement has been worked out after much thought and experience with the acid.

Dilute Sulfuric Acid

When 60° Baumé sulfuric acid is added to water, the mixture becomes a true acid and is no longer characterized by burning. When you prepare a dilute solution of 5 to 10%, you must always add acid, in small increments, to a large volume of water. If the solution temperature gets too hot to the touch, you will have to wait until it cools before adding additional acid.

Under no circumstances should you add water to a carboy of 60° Baumé sulfuric acid. The reaction would be so intense that acid could splash onto your eyes and clothing

At this concentration, 5 to 10%, sulfuric acid will cause a great deal of discomfort if it is splashed onto eyes or clothing. On skin it will produce an immediate stinging sensation, and in the eyes a similar but much more severe sensation will be experienced.

Reagent Strength Acid This is the acid normally found in test kits. Some kits specify the strength in chemical terms, for example, 0.1N, or $\frac{1}{10}N$, and others just list *dilute sulfuric acid*. The nomenclature is not too important for the layperson but is for the chemist. As a rule of thumb, 0.1N, or $\frac{1}{10}N$, is about 1% solution of 60° Baumé sulfuric acid in distilled water. Another commonly used strength is 0.02N of $N/50$, which is about 0.2% 60° Baumé acid in distilled water.

At this strength it should pose no significant problem to the user; however, *if it gets into the eyes, it will sting!*

Handling Sulfuric Acid Spills The reagent-strength acid can just be wiped up with a paper towel and disposed of in the wastepaper basket. At other strengths, you should add to it a mixture of sawdust and soda ash and shovel the whole mess into a plastic garbage bag. Disposal of the garbage bag should not pose a problem, but you should alert the garbage collectors as to its contents. For the final cleanup, wipe down the spilled area with a 5% solution of soda ash and then flush with city or tap water.

Quaternary Ammonium Compounds

This group of compounds is used for algae control, and perhaps some bacterial control as well, in cooling towers and other recirculating water systems. Being soaps in nature, these compounds function by removing the first line of defense of cells, the lipid or fat layer. When this happens, the permeability of the cell wall is reduced, and so water can enter with ease, thus destroying the cell. This explanation is rather simplistic, but it serves the purpose.

As to safety, if you look at labels listing a quaternary ammonium compound, you are confronted with unpronounceable multisyllable words, letters, and numbers, such as n-alkyl (60% C_{14}, 30% C_{16}, 5% C_{12}, 5% C_{18}) dimethylbenzylammonium chloride. Why this is done is beyond the author's comprehension, since no one ever looks at these labels other than to read the precautions anyway. For the sake of simplicity, this drawing represents such a compound:

$$R_1 - \underset{\underset{R_4}{|}}{\overset{\overset{R_2}{|}}{N}} - R_3\text{-chloride}$$

Under the right conditions and with the correct R's, *this group can form highly carcinogenic N-nitroso compounds.* Note that this is a *possibility;* the reaction may not proceed, but it could. Here is what an N-nitroso compound looks like:

$$N\text{-nitroso compound:} \qquad \overset{R}{\underset{R}{>}} N - N = O$$

Since these are algaecides, it would be wise to heed the warnings on the labels. Use only that quantity required to do a specific job — no more, no less. Since these compounds will work for only a short time (i.e., certain portions of the unicellar population will become immune and overwhelm the system), you will be forced to switch to an alternative.

Thiocarbamates

These chemicals comprise a direct poison group that functions by penetrating the cell wall and short-circuiting the life process within. *This group, as pictured below, also seems capable of producing N-nitroso compounds.*

$$\overset{R}{\underset{R}{>}} N = C \overset{S}{\underset{S}{<}} \text{metal} + NO_2 = \overset{R}{\underset{R}{>}} N - N = O$$

Since this is an algaecide, the same precautions apply to this group as to the quaternary compounds. A certain portion of the unicellular population will become immune to the action of thiocarbamates, and you will be forced to seek an alternative.

Halogens

The halogens are the compounds of last resort in algae control, to be used only when all others fail. They include chlorine gas and all the liquid and solid preparations capable of releasing chlorine and bromine into the system, to form the halogen acid that kills the unicellular life forms. The danger involved in handling these compounds is well known. However, at times one is reminded when a container of powdered or brick chlorine compound is opened after being left standing in the hot sun. Those who handle halogen compounds must remember that under the right conditions these chemicals can injure or kill people.

2-Mercaptobenzothiazole

This compound, also known as mercaptobenzothiazole (MBT), is used as a corrosion inhibitor in open and closed recirculating systems. Since it is fatal to goldfish at a concentration of 2.0 ppm in 48 h, *a heavy spill into a stream or lake could take its toll on fish.* If MBT is used as a corrosion inhibitor in a closed hot-water heating system and the system lacks back-flow preventers, the system water can enter the domestic water system and be ingested by the unwary. Since MBT is used in conjunction with nitrites and borates and thus is colorless (unless a very strong dye is added to the formulated product), its use could result in contamination of a system with a double dose of potentially hazardous products.

STEAM BOILERS

Sulfites

Sulfites are used as an oxygen scavenger, with or without added catalysts, to accelerate the reaction rate. Sulfites are also added to some foods to maintain the appearance of freshness, and *when ingested, they are known to trigger respiratory upset in asthmatics.* Consumer groups want to ban their use in foods or at least force food manufacturers to list them as ingredients. *Maintenance personnel known to suffer with asthma should avoid handling boiler compounds containing sulfites.* The OSHA material safety data sheet should list any such product, if the formula contains it, under "Formula."

Hydrazine

Hydrazine is used as an oxygen scavenger, in lieu of sulfites, when added solids are not desirable. Under pressure and in the presence of oxygen, hydrazine decomposes into water and nitrogen gas. This product should be handled with great care since it can enter the body through the skin. *It has been shown to be carcinogenic in rats and mice, but tests on other animals have failed to produce malignant tumors.* The threshold limit value (TLV) exposure to the compound in any 8-h shift of a 40-h week is as follows:

Country	TLV, concentration
U.S.S.R.	0.08 ppm
United States	1.0 ppm
Germany	0.1 ppm

It is interesting that the United States allows its workers to be exposed to hydrazine at a level 12.5 times greater than that allowed by the Soviet Union.

Morpholine

This product is a neutralizing amine for steam boilers. It is also known as tetrahydro-1,4-oxime and diethylenimide oxide, and its configuration is as follows:

$$OCH_2CH_2NHCH_2CH_2$$

The lethal dose (LD_{50}) of a product is that level which is expected to kill 50 percent of the population of test animals; i.e., 50 percent of the test animals will die at this dosage. Some interesting LD_{50}'s for morpholine are as follows:

Guinea pig   Oral LD_{50} = 0.9 g/kg of body weight

Rat   Oral LD_{50} = 1.0 to 1.6 g/kg of body weight

For long-term exposure we notice the following effects after the animals were fed the dosage daily for 30 days:

Rat   Oral dosage killed 19/20 @ 0.8 g/kg of body weight

Guinea pig   16/20 @ 0.45 g/kg of body weight

    3/20 @ 0.04 g/kg of body weight

The TLVs again show an interesting contrast between the Soviet Union and the United States in terms of allowed exposure:

Country	Concentration	TLV
U.S.S.R.	0.15 ppm	0.5 mg/m³ of skin
United States	20.00 ppm	1.05 mg/m³ of skin

Since morpholine is a volatile amine and since the TLVs are noteworthy, all necessary steps should be taken to limit human exposure to this compound. Obviously it should not be employed where steam is used for humidification or food preparation. The humidification may or may not be intentional; e.g., defective steam traps could fill the boiler room atmosphere with heavy concentrations of the product.

Cyclohexylamine

Another neutralizing amine used in steam boilers for protection of the return lines is cyclohexylamine. It is also known as hexahydroaniline and aminocyclohexane and may appear as follows:

$C_6H_{11}NH_2$

The TLVs are as follows:

Country	Concentration	TLV
U.S.S.R.	0.25 ppm	1.0 mg/m³ of skin
United States	10.0 ppm	10.0 mg/m³ of skin (tentative 1972)

The same care should be taken in the use of this product as was urged for morpholine. *The fact that TLVs are set indicates a certain risk in using this product.*

Diethylethanolamine

Diethylethanolamine is yet another neutralizing amine for steam use. This product is also known as 2-diethylethanolamine, β-diethylaminoethylalcohol, 2-hydroxytriethylamine, and DEEA. Its configuration is as follows:

$$CH_3CH_2 \diagdown$$
$$N-CH_2CH_2OH$$
$$CH_3CH_2 \diagup$$

The TLVs are again quite contrasting:

Country	Concentration	TLV
U.S.S.R.	1.0 ppm	5 mg/m³ of skin
United States	10.0 ppm	50 mg/m³ of skin

This product was reported to have coated and perhaps damaged artworks at the H. F. Johnson Museum at Cornell University.[7] Employees also reported feeling ill during the time the amine was being vented in the library. This problem may have led to a closer scrutiny in the use of amines in return-line systems. A report done for the U.S. Navy suggests that the use of neutralizing amines and nitrite boiler compounds could lead to the formation of volatile nitrosamines, which are carcinogenic compounds. Thus some reputable water treatment firms no longer propose that program.

Filming Amines

Octadecyl-, hexadecyl-, and dioctadecylamines are filming amines, in contrast to neutralizing amines. Their configuration is as follows:

$$CH_3(CH_2)_{17}-N\Big\langle\begin{array}{c}H\\[1mm]H\end{array}$$

Because of their high boiling points — 232°F (111°C) at 32 mmHg for octadecylamine — these products are not easily vaporized and spread into the atmosphere. The difficulty in maintaining the protective film also hinders their widespread use, as does the oily mess they leave on valves, stems, and other equipment.

Sodium Hydroxide

This product is used for pH control in steam boilers, and it is also known as lye and caustic soda. It can be obtained as dry flakes or pellets and is also available in liquid form in various concentrations. As a boiler water additive, it may be supplied with various ingredients to accomplish something other than just pH control. *Sodium hydroxide, no matter how disguised, must be handled with extreme caution because it can cause blindness if it comes into contact with the eyes.* It will react with grease and oil, animal or vegetable, to form a soap — thus the customary soapy feeling you get when you try to cleanse your hands of the material.

CLEANING COMPOUNDS

Hydrochloric Acid

This product is used for descaling operations, and it is also known as muriatic acid and hydrogen chloride. The term "hydrogen chloride" would be incorrect since it is reserved for the gas, not the acid. The acid is nothing more than hydrogen chloride gas, HCl, dissolved in water to form H^+ and Cl^-. In cleaning operations the H^+ reacts with the lime, $CaCO_3$, to convert the CO_3^{2-} to carbon dioxide, a gas.

As with any acid, you should exercise caution in its use and be prepared to clean up spills. Goggles and protective clothing are a must. If you descale a boiler by using this acid, make sure there is adequate ventilation because hydrogen chloride gas will be given off when the boiler is warmed up, to speed up the reaction and thus cut down labor hours. *The hydrogen chloride gas will also adversely affect those with upper respiratory problems, such as asthma and emphysema.*

Should you buy hydrochloric acid with an inhibitor (a chemical added to protect the bare metal exposed during the cleaning process), request a copy of the OSHA safety data sheet for each inhibitor used. Do not assume that the acid is the dangerous component of the formulation, because some inhibitors, if allowed to enter the body through contact or the respiratory system, can cause serious health problems.

Formulated Products for Cleaning HVAC System Components

Users of these products are at a distinct disadvantage because important information is sometimes deleted from or hidden in the OSHA sheet in order to make them more appealing. Let us consider the products in a generic manner and learn what to look for:

Surface Cleaners These products contain a solvent, surfactant, and water with a trace of color and perhaps a little perfume of sorts to hide any odor. The OSHA sheet should discuss the nature of each component (e.g., which is the solvent?) and whether it adversely affects a person using it to clean a piece of equipment in a confined area.

Grease Removers *These products may contain caustic soda, or sodium hydroxide,* which has already been discussed. They may contain organic grease cutters which may or may not be harmful. Although the product itself may be safe, it could liberate harmful gases when used to clean certain areas. The user should be made aware of the nature of the product and contraindications to its use.

Coil Cleaners These products may be alkaline or acidic, depending on their formulation. *If a coil cleaner is alkaline, it may contain caustic soda, which will attack aluminum.* (It will also liberate hydrogen gas, which in itself is not dangerous since the gas will not be generated in sufficient quantities to cause problems.) Thus, besides the dangers already discussed in relation to caustic soda, *constant use of the product will destroy aluminum fins and coils.*

If the coil cleaner is acidic, it may contain one or more acids, mineral as well as organic, e.g., sulfuric and sulfamic acids. The formulator may try to hide the fact that a product contains an acid by using other than correct terminology: "Hydrochloric acid" and "muriatic acid" are understood terms, but the formulator may list hydrogen chloride as an ingredient. *The constant use of acid on aluminum fins will certainly shorten the life of the coil.*

Another important consideration is the by-products of reaction. The product, as sold, may be as safe as milk, but it could cause the evolution of dangerous products of reaction. If a product contains hydrochloric acid and hydrogen chloride is listed on the OSHA data sheet as a hazardous decomposition product, then you can be sure that the hydrogen chloride gas will react with the moisture in the lungs and in the eyes to form hydrochloric acid. *The same is true of products that list hydrogen fluoride as a decomposition product, or product of reaction. When hydrogen fluoride is in contact with water or moisture, it will form hydrofluoric acid, which attacks glass and could lead to sight loss* (the gas turns to acid when in contact with naturally occurring moisture in the eyes).

It is impossible, in a work of this nature, to list all the dangerous properties of chemicals and compounds used in the HVAC industry. Even if all the presently used organic chemicals could be listed, the nature of organic chemistry is such that the information would be outdated by publication date. To help make your library complete, the author highly recommends the following additions:

Karel Verschueren, *Handbook of Environmental Data on Organic Chemicals,* Van Nostrand Reinhold, New York, 1977.

N. Irving Sax, *Dangerous Properties of Industrial Materials,* 5th ed., Van Nostrand Reinhold, New York, 1979.

Note that the words "EPA registered and approved by the Department of Conservation" in a given state do not mean, although they do hint, that the water treatment company or approving agency has tested the product and found it safe to use. *More often than not, the user is the ultimate test animal.* Unless it is proved otherwise, you would be prudent to treat every drum of chemicals as if it were extremely hazardous material. The following steps should be taken:

1. *Ask for an OSHA safety data sheet, and make certain all chemical handlers read and understand it.*

2. *Inquire as to the nature of each ingredient.*

3. *Request a statement, in writing on company letterhead, that each ingredient and the combination of ingredients, used in accordance with directions, as well as the products of reaction are not harmful to the user or the environment.*

4. *Do not mix or add chemicals contrary to instructions, for then no one can*

predict the outcome. The reaction products could be harmless, harmful, or fatal!

5. *If different compounds, or treatments, are to be added to the system, try to do it on alternate days. For example, add product A on Monday and product B on Wednesday.*

6. *Handlers should wear gloves and wash their hands after using chemicals, even the chemicals in test kits.*

7. *"The nose knows." So if it smells bad, it undoubtedly is; use caution.*

By now you have gathered sufficient information from this chapter to purchase and use water treatment chemicals in a wise and safe manner. Remember, a verbal guarantee that a product is as safe as milk, or as dangerous as cyanide, is not worth the paper it is printed on. Get it in writing!

EIGHT

Water Treatment Specifications for Existing Buildings

ESTABLISHING VIABLE WATER TREATMENT SPECIFICATIONS

For an existing building, you can use this book to upgrade the chemical feed system. If you learn that a present practice is not in keeping with a good program, then do whatever is necessary to correct it. Public and private educational facilities are known for their lack of attention, intentional or unintentional, to the water treatment requirements of their systems. However, they are not alone in this respect.

In New York State, when a public institution exceeds a certain level of expense for water treatment, it must begin to think in terms of open, competitive bidding in order to stay within the requirements of the law. The following water treatment specifications should help you get the most for your water treatment dollar. They can be employed by any water treatment user, with the limiting factor being the potential money available to the water treatment firms.

WATER TREATMENT SPECIFICATIONS

These specifications can be used to request bids for supplying all necessary chemicals for systems listed:

Cooling Towers

Chemicals shall be supplied for cleaning the systems prior to startup and for the control of algae, alkalinity, corrosion, dirt, and scale for a period of ___[list time period]__ from date of startup.

Corrosion shall be held to less than 5.0 mils/yr and shall be measured by the use of corrosion coupons placed in the cooling tower pan* or in a by-pass arrangement.†

Alkalinity shall be controlled by adding sulfuric acid or sulfamic acid in such quantities as are required to maintain the alkalinity within 350 to 400 ppm, expressed as calcium carbonate, $CaCO_3$. The chemical feed equipment for adding

---

* NOTE: Not *on* the pan, but suitably arranged to keep the coupon under running water at all times.

† See Fig. 28.

acid shall be as shown in the accompanying drawing* and shall be supplied by the bidder.

Installation shall be by [list who is to install] .

Scale control shall be effected via the addition of suitable corrosion inhibitors and bleed-off. At contract's end the condenser (or tubes of HX in the evaporative condenser) shall be opened and inspected. If scale, as insoluble salts of calcium and magnesium, should be found, it shall be cleaned at the expense of the water treatment company if all recommendations, in writing, have been followed.

Algae and the growth of unicellular organisms shall be controlled via the addition of EPA-approved biocides on a schedule as determined by the water treatment company. The water treatment company shall include in the bid a sufficient sum to allow for the testing of the water for bacterial activity† by an independent laboratory. The use of dip slides for bacterial testing will not be acceptable. Should algae or hemolytic activity develop during the contract period, the staff having followed all written instructions and maintained an appropriate log, it shall be the responsibility of the water treatment company to supply all chemicals and labor for cleaning. Any area that is not constantly flooded with water, such as slats and fan blades, shall not fall under the cleaning requirement.

Closed System

Chemicals supplied shall control corrosion down to 5.0 mils/yr or less. The chemicals shall not undergo decomposition or bacterial reversion while in the system. Should this occur, it shall be the responsibility of the water treatment company to supply chemicals and labor to correct conditions immediately and prevent a recurrence.

Steam Boilers and Return-Line System (Low-Pressure Units)

Chemicals shall be supplied for cleaning the steam boiler(s) of any waterside incrustation if, on inspection, it is found to contain same. After cleaning the units, which shall be done by [list who will do the cleaning] *, and prior to startup, a sufficient quantity of boiler treatment shall be added to prevent the formation of waterside incrustations and to hold corrosion to less than 5.0 mils/yr‡ in the boiler proper.*

At the end of the contract period, the boilers shall be opened and inspected. Should any scale or waterside incrustation be found, it shall be the responsibility of the water treatment company to clean same, at its own expense. The inability of the owner to keep the steam loss at a low level (i.e., to fix steam traps when cycles of concentration are found to be climbing and this is brought to the attention of the owner) shall free the water treatment company of this cleaning re-

---

* See Fig. 56.

† Testing should be done on a monthly basis during the contract period, by means of four dishes of Trypticase soy agar with 10 percent sheep's blood.

‡ This is really too generous!

quirement. However, the water treatment company must warn the owner, in writing, about the consequences of such a steam loss.

For return-line corrosion, a sufficient quantity of an approved (FDA, EPA, and OSHA) volatile corrosion inhibitor shall be added in such quantities as to afford adequate corrosion protection. The corrosion rate shall be held to less than that of an untreated system. A corrosion coupon may be used for 20 days to ascertain the level of no protection. Under no circumstances shall a product containing a formulated mix of boiler treatment and amines be supplied.* The return-line and boiler treatments shall be separate and distinct treatments in their own containers. No deviations will be accepted.

Steam Boilers and Return-Line System (Medium- to High-Pressure Units)†

Domestic Water Systems

A sufficient quantity of a health-department-approved, food-grade corrosion and scale inhibitor shall be supplied to protect the domestic water systems against scale and corrosion.‡ The corrosion shall be kept to less than 5 mils/yr in the __[list]__ system. The domestic heat exchanger shall show no signs of scale or incrustation at contract's end. Should the unit show signs of scale or incrustations, it shall be cleaned by the water treatment company at its own expense.

Chemical Feed Systems

It shall be the responsibility of the water treatment company to ascertain fitness of chemical feed systems in use. Should the water treatment company find that alterations are required, it shall be the responsibility of the bidder to include the required items as a separate part of the bid. The owners will review the justification for the alterations and may accept or reject same; however, the water treatment company will be held to bid on only that which is specified. The additional materials will not be considered as part of the bid per the specifications.

Test Kits

All necessary test kits shall be supplied by the bidder, as well as extra reagents, for the duration of the contract. Present test kits may be used, if suitable.

---

* Read Chap. 7, "Chemical Safety," in regard to the use of volatile amines for protection against the corrosion of return lines.

† Use the same basic specification as outlined in Chap. 5. But be guided by what was said just above.

‡ At this stage one should have had the means for adding the treatment already installed.

Cancellation of Contract*

System	Specifics
Cooling tower(s)	Tonnage(s) _____
	Pump, gal/min _____
Refrigeration equipment	
Make	_____
Model	_____
Type (absorber, reciprocating, . . .)	_____
Tonnage	_____
Closed system(s)	
Chilled	Pump, gal/min _____
Hot water	Pump, gal/min _____
Heat exchanger	Capacity, gal/min _____
Steam boilers	
Make	_____
Horsepower	_____
Steam, lb/h	_____
Pressure	_____
Use (process or heating)	_____
Return condensate, %	_____
Domestic water system — Estimated use, gal/day	
Cold	_____
Hot	_____

The above list may not be accurate. It shall be the responsibility of the water treatment company to ascertain correctness of figures and bid on same.

---

* Insert appropriate comment here. You may wish to cancel the contract if the representative of the water treatment company does not exude confidence in the approach to the program. The realities of the marketplace are such that water treatment firms may be forced to hire individuals with limited, if any, expertise in the water treatment field just to maintain representation in a given area. On the other hand, you are interested in someone who knows water treatment and the full ramifications of deviations from the norm.

Quality Assurance

The water treatment firm shall have been regularly engaged in the water treatment of __[list]__ years. The field engineer shall have no less than __[list]__ years' experience in actual, direct charge of onsite testing of similar systems. Further, the field engineer shall be a graduate chemist, A.A.S. or B.S., from an ACS-accredited institution* and shall make inspection visits no less than monthly. The field engineer shall be available, on call, within forty-eight (48) hours after being contacted to correct any emergency situation such as, but not limited to:

1. *Chemical imbalance of system*

2. *Chemical spills*

3. *Bacterial problems*

4. *Questions about OSHA guidelines*

The field engineer shall also be available for training purposes on an on-call basis.

Upon request, the water treatment company shall submit a list of at least __[list number]__ installations of similar capacity in __[insert your home state or a specific geographical area]__ which have been successfully treated for a period of __[list number]__ years immediately preceding the award of this contract.

It shall be the responsibility of the field engineer to test all systems and make recommendations, in writing, to the owners to correct any condition found unacceptable.

Quotation

Cooling tower	
Scale and corrosion inhibitor	*Price/lb $_____*
Amount per one thousand (1000) gal of make-up	_____
Algaecide	*Price/lb $_____*
Amount per one thousand (1000) gal of make-up	_____
Coagulant	*Price/lb $_____*
Amount per one thousand (1000) gal of make-up	_____
Sulfuric acid	*Price/lb $_____*
Amount per one thousand (1000) gal of make-up	_____
Closed chilled-water system	
Corrosion inhibitor	*Price/lb $_____*
Amount per one thousand (1000) gal of make-up	_____
Closed hot water	
Corrosion inhibitor	*Price/lb $_____*
Amount per one thousand (1000) gal of make-up	_____

(Continued)

---

* Would you want someone with a degree in music telling you how to maintain your system, or would you prefer a trained chemist?

Steam boiler(s)
   Boiler treatment   *Price/lb $*_____
   Amount per one thousand (1000) gal of make-up   _____
Domestic water systems
   Treatment   *Price/lb $*_____
   Amount per one thousand (1000) gal of make-up   _____
Preseason sanitizing agent for all open systems to
make systems sanitary prior to startup
   Sanitizing agent   *Price/lb $*_____
   Amount per one thousand (1000) gal of make-up   _____
   Total, F.O.B. delivered   *$*_____
   Terms _____
   Discount for prompt payment _____

Upon award of this contract, the bidder shall submit a data sheet and an OSHA safety data sheet for each product. No system shall be treated with any chemical or compound for which there is no test procedure, except for weekly slug chemicals. The closed loops shall not be treated with any chemical or compound that can undergo bacterial attack or reversion while in the system. Any exceptions to the specifications shall be written and submitted along with the bid. Any questions on the bid shall be brought to _____ *at* _____ *ext.* _____ *during the hours of* _____ *and* _____ *.*

These specifications can be used for any building with slight alterations. Note that we have not told the water treatment company what to use. In keeping with the spirit of this work, *we are requesting performance!* Perhaps we are asking too much; but that decision must rest with the person spending the funds for the program.

"GADGETS"

No work on water treatment is complete without a section on "gadgets." The author's file on the subject is presently about 2 in thick and growing! In this day and age, with all that is known, still plant operators fall for the sales pitch of the purveyors of these worthless pieces of metal. *Do they work?* If I had been talked into spending company funds for a gadget (some units cost in excess of $5000 depending on pipe size), of course I would have nothing but praise for it and it would work — if only as a brace to hold up a shelf. *How many would openly admit to their superiors that they had squandered company funds on a worthless piece of metal?* These units have been tested, retested, exposed, and written about since 1865, when a lightening rod in a boiler used the magical powers of electricity to protect it. Hope springs eternal (P. T. Barnum had a better phrase), but even so, here is a list of references for you to consult. But having read all these works, and more, the author can advise you ahead of time not to throw away money by purchasing such items of "water treatment" garbage.

R. Eliassen and H. H. Uhlig, "So-Called Electrical and Catalytical Treatment of Boiler Water," *Journal of the American Water Works Association,* vol. 44, no. 7, July 1952.

R. Eliassen, R. T. Skrinde, and W. B. Davis, "Experimental Perform-ance of Miracle Water Conditioners," *Journal of the American Water Works Association,* vol. 50, no. 10, Oct. 1958.

"Federal Trade Commission Decision on Evis Water Conditioner Claims," *Journal of the American Water Works Association,* vol. 51, no. 6, June 1959.

H. P. Godard, "Watch Out for Wonderous Water Treatment Witch-craft," *Materials Performance,* vol. 13, no. 4, Apr. 1974.

Milton Meckler, "Electrostatic Descaler Testing: An Evaluation," *Heating, Piping and Air Conditioning,* Aug. 1974.

References

1. Sheppard T. Powell, *Water Conditioning for Industry,* McGraw-Hill Book Company, New York, 1954.

2. Rolatrol is a registered trademark of ITT, B&G Inc.

3. Phenophthalein indicator, available from water treatment firms, goes from colorless to pink at a pH of 8.3. The higher the pH, the deeper the color. Since you are adding an alkaline compound to the system water, phenolphthalein will turn pink or red, assuming the raw water has a normal pH of less than 8.3. When a sample fails to turn pink, you can assume the system has been adequately flushed. If the city water supply has had the pH artificially increased, e.g., Utica, NY, you must use an alternative means for ascertaining the extent of flushing; see note 5 below.

4. Sidney Sussman, "Causes and Cures of Mechanical Shaft Seal Failures in Water Pumps," *Heating, Piping and Air Conditioning,* Sept. 1963.

5. The specifying engineer should consider that 7 working days is not too short a time to check system water for completion of cleaning. Excellent test kits are available; see the Product List for one to check, on the spot, for cleaning compound residual. For example, to check for phosphates, there is the Hach No. 1473-01 test kit, while for silicas, there is the Hach No. 1478-01 test kit. There is no reason why the system cannot be tested immediately after final flushing. It is not unreasonable to expect a water treatment firm, or consultant, to be prepared to perform a specific job.

6. Frank Rosa, "Biocides — Algaecides. I. Do They Protect You?" *National Engineer,* Mar. 1982; "Biocides — Algaecides. II. How Much Protection?" *National Engineer,* Apr. 1982; "Biocides — Algaecides. III. How Much Protection?" *National Engineer,* May 1982.

7. Wayne Biddle, "Art . . . Coated by Chemical Used in Steam Lines," *The New York Times,* July 29, 1983.

8. R. H. Marks, "Water Treatment: Part One," *Power,* Dec. 1958, special report; "Water Treatment: Part Two," *Power,* Mar. 1959, special report.

9. *Handbook of Air-Conditioning System Design,* McGraw-Hill Book Company, New York, 1965.

10. Richard T. Blake, "Correct Water Treatment Can Save Energy," *Building Systems Design Magazine,* Apr./May 1977.

11. Frank Rosa, "Fan Inefficiency May Be due to Deposit Build-up," *National Engineer,* July 1982.

12. W. A. Keilbaugh and F. J. Pocock, "Pointers on the Care of Low Pressure Steam Steel Boilers," *Heating, Piping and Air Conditioning,* Feb. 1962. Although while this report suggests a correlation between pitting and high chloride levels, the author failed to substantiate it via field experience. Low-pressure steam steel boilers with chlorides in excess of 600 ppm, expressed as sodium chloride, treated with chromates have been operating in excess of 15 years with no significant pitting. This is not to imply that the chloride content of the make-up water should be ignored. It is a factor that must be considered, especially if a nonchromate treatment will be used.

13. Frank Rosa and James J. Glass, "Design of Steam Boiler Installations May Be Contributing Factor to Inefficiency," *National Engineer,* May 1980.

14. Frank Rosa, "Boiler Flooding Problems," *Heating, Piping and Air Conditioning,* Oct. 1983.

15. United Nations World Health Organization, International Agency for Research on Cancer, *IARC Monographs,* 1974.

Product Directory

Backflow Preventers

Lawler ITT
3500 North Spaulding Ave., Chicago, IL 60618
(312) 463-0222

Boiler Blowdown — Continuous

Dias Incorporated
2327 Winters Dr., Kalamazoo, MI 49002
(616) 344-1008

Morr Control, Inc.
P.O. Box 632, Muskogee, OK 74401
(918) 683-0238

Speciality Valve and Controls
P.O. Box 10, Fairview, PA 16415

Boiler Water — Sample Coolers

Crane Company, Cochrane Division
P.O. Box 946, Wall Street Station, New York, NY 10268
(212) 775-1395

Madden Corporation
22 S. Washington St., Park Ridge, IL 60068
(312) 696-2303

Conductance Meters and Controllers

Cambridge Scientific Industries
101 Virginia Ave., Cambridge, MD 21613

Chemtrix, Incorporated
163 S.W. Freeman, Hillsboro, OR 97123
(503) 648-0762

Clack Corporation
P.O. Box 500, Windsor, WI 53598
(608) 251-3010

Morr Control, Inc.*

Myron L. Company
1133 Second St., Encinitas, CA 92024
(714) 753-6215

---

* The addresses and phone numbers of starred entries appear elsewhere (earlier) in the list.

195

Corrosion Monitors

Magna Corporation
11808 South Bloomfield Ave., Santa Fe Springs, CA 90670
(213) 863-4781

Petrolite Corporation
5455 Old Spanish Trail, Houston, TX 77001
(713) 926-7431

Questronics
10346 Mississippi Ave., Los Angeles, CA 90025
(213) 553-2679

Rohrback Instruments
11861 E. Telegraph Rd., Santa Fe Springs, CA 90670-9915
(213) 949-0123

Filters

AMF/Cuno
400 Research Parkway, Meriden, CT 06450

Bird Machine Company, Inc.
100 Neponset St., So. Walpole, MA 02071
(617) 668-0400

Chemworks
289 Mt. Hope Ave., Dover, NJ 07801
(201) 328-0548

Fluid Filtration Systems
3301 Gilman Rd., El Monte, CA 91732
(213) 443-4211

Inlay Incorporated
P.O. Box 471, Trimmer RD., Califon, NJ 07830

Laval Separator Company
1899 North Helm, Fresno, CA 93727
(209) 255-1601

Ronningen-Petter
P.O. Box. 188, Portage, MI 49081
(616) 323-1313

Smith-Koch, Inc.
2536 So. 59th St., Philadelphia, PA 19143
(215) 726-7100

Flowmeters

Aalborg Instruments & Controls, Inc.
57 Regina Rd., Monsey, NY 10952
(914) 352-3171

Blue White Industries
14931 Chestnut St., Westminster, CA 92683
(714) 893-8529

Clack Corporation*

Delaval Turbin, Inc.
Farmington, CT 06032
(203) 677-1311

Hedland
2200 South St., Racine, WI 53404
(414) 639-6770

Matheson Instruments
418-3 Caradean Dr., Horsham, PA 19044
(215) 674-0686

RCM Industries
P.O. Box 351, Orinda, CA 94563
(415) 687-8363

Schutte & Koerting Company
Cornwells Heights, Bucks County, PA 19020

The W. A. Kates Company
P.O. Box 627, Deerfield, IL 60015

Magnetic Traps

Indiana General, Magnetic Equipment Division
6001 S. General Ave., Cudahy, WI 53110
(414) 482-1500

Hayward Industrial Products, Inc.
900 Fairmont Ave., Elizabeth, NJ 07207
(201) 351-5400

S. G. Frantz Company, Inc.
31 East Darrah Lane, Trenton, NJ 08606
(609) 882-7100

pH Controllers

Analytical Measurements, Inc.
31 Willow St., Chatman, NJ 07928
(201) 273-7500

Chemtrix*

Delta Scientific Corporation
120 E. Hoffman Ave., Lindenhurst, NY 11757
(516) 884-4422

Devon Products, Inc.
7321 N. Figueroa St., Los Angeles, CA 90041
(213) 257-7585

Horiba Instruments, Inc.
1021 Duryea Ave., Santa Ana, CA 92705
(714) 540-7874

Kernco Instruments Company, Inc.
708 N. Piedras St., El Paso, TX 79903
(915) 566-9651

Leeds & Northrup
SumneyTown Pike, North Wales, PA 19454

Questronics*

Pumps, Acid (for cleaning)

March Manufacturing Company, Inc.
1819 Pickwick Ave., Glenview, IL 60025
(312) 729-5300

Multi-Duti Manufacturing, Inc.
1414 Live Oak Ave., Baldwin Park, CA 91706
(213) 357-5091

Proven Pumps Corporation
1440 North Spring St., Los Angeles, CA 90012
(213) 223-3101

Sethco Manufacturing Corporation
One Bennington Ave., Freeport, NY 11520
(516) 623-4220

Pumps, Chemical Feed

Blue White Industrues*

Chem-Tech International*

Gorman-Rupp Industries
Bellville, OH 44813
(419) 886-3001

Hydroflo Corporation
Plumsteadville, PA 18949
(215) 249-9090

Interpace Corporation
77 Ridgeland Rd., Rochester, NY 14623
(716) 424-5600

Liquid Metronics, Inc.
19 Craig Road, Acton, MA 01720

Neptune Chemical Pump Company
Lansdale, PA. 19446
(215) 699-8701

Precision Chemical Feed Pumps
Waltham Chemical Pump Company
1396 Main St., Waltham, MA 02154

Strip Chart Recorder

Gulton Industries, Inc.
Gulton Industries Park, E. Greenwich, RI 02818
(401) 884-6800

Tank Liner (for Corroded Tank)

Flexi-Liner Corporation
44 S. Raymond Ave., Pasadena, CA 91102
(213) 796-3117

Tanks

Brumley Equipment, Inc.
340 Main St., Springfield, MA 01105
(413) 736-4587

Chem-Tainer Industries
361 Neptune Ave., N. Babylon, NY 11704
(516) 661-8300

Terracon Corporation
1396 Main St., Waltham, MA 02154

Test Kits

Chemetrics, Incorporated
Warrenton, VA 22186
(703) 347-7660

Chem-Tech International
    Merrimack & South Union St., Lawrence, MA 01843
    (617) 685-4301

Hach Manufacturing Company
    P.O. Box 389, Loveland, CO 80539
    (800) 525-5940

Hellige Incorporated
    877 Stewart Ave., Garden City, NY 11530
    (516) 222-0300

LaMotte Chemical Company
    Chestertown, MD 21620
    (301) 778-3100

Timers

Eagle Signal Industrial Controls
    736 Federal St., Davenport, IA 52803

Potter & Brumfield
    Princeton, IN 47671
    (812) 386-1000

The Thrush Group
    67–69 Albany St., Cazenovia, NY 13035
    (315) 655-8476

Valves

Hancock, Div. Dresser Industries, Inc.
    245 Park Ave., New York, NY 10017
    (212) 697-2800

Lunkenheimer
    Cincinnati, OH 45214

Hayward Industrial Products, Inc.*

Water Meters

Carlon Meter Company
    715 Robbins Ave., Grand Haven, MI 49417
    (616) 842-0420

Kent Meter Sales, Incorporated
    7 E. Silver Springs Blvd., Ocala, FL 32670
    (904) 732-4670

Water Softeners

Barnstead
    225 Rivermore St., Boston, MA 02132
    (617) 327-1600

Bruner Corporation
    4767 N. 32d St., Milwaukee, WI 53209

Culligan
    One Culligan Parkway, Northbrook, IL 60062
    (312) 498-2000

Ion Exchange Products, Inc.
    4500 No. Creek St., Chicago, IL 60640
    (312) 784-4100

Permutit Company, Inc.
    East 49 Midland Avenue, Paramus, NJ 07652

Water Treatment Companies

The author has not included a list of water treatment companies for the following reasons:

1. The list would be too long, for there are over 100 water treatment companies.

2. Since water treatment chemistry is not an exotic art, any firm with a desire to expand its product line from swimming pool chemicals to floor waxes can get into water treatment.

3. One has no way of knowing the quality of the backup, i.e., laboratory service, being provided.

4. All things being equal, the reputation of a particular water treatment firm depends on its field representatives. A firm may have an excellent reputation in one state but, because of poor representation, a bad reputation in another state.

5. The qualifications of representatives are of paramount importance. The representative with a strong background in chemistry and biology will prove to be of great value to the client. If the representative is lacking in those areas, the client is paying too much for the goods and services regardless of the actual price. But if the list of water treatment firms would be too long, the list of representatives would be even longer.

You may well make longevity the primary consideration in choosing a water treatment company; however, this is not to say that a young firm should be ignored. It is a difficult decision that merits a great deal of thought, as did the choice of equipment when you were in the market for it.

Bibliography

It is recommended that you obtain the following books and make them part of your permanent library. Even though the potential combinations in organic chemistry alone could exceed the number of stars in the universe, the principles governing scale formation, corrosion, algae, and bacterial control are known, and corrective measures are fairly well established.

Subscription to: *Heating/Piping/Air Conditioning*
                Reinhold Publishing
                P. O. Box 95759, Cleveland, OH 44101

Books:

Handbook of Air-Conditioning System Design, 1965.

Frank N. Speller, *Corrosion, Causes and Prevention,* 1951.

Sheppard T. Powell, *Water Conditioning for Industry,* 1954.

Richard T. Blake, *Water Treatment for HVAC and Potable Water Systems,* 1980.

All the above from: McGraw-Hill Book Company
                   1221 Avenue of Americas, New York, NY 10020

Karel Verschueren, *Handbook of Environmental Data on Organic Chemicals,* 1977.

N. Irving Sax, *Dangerous Properties of Industrial Materials,* 5th ed., 1979.

Above from: Van Nostrand Reinhold
            135 W. 50th St., New York, NY 10020

Richard E. LaFond, *Cancer, The Outlaw Cell,* 1978.

Above from: American Chemical Society
            1155 16th St., N.W., Washington, DC 20036

United Nations, World Health Organization, *IARC Monographs on the Evaluation of Carcinogenic Risk of Chemicals to Man,* vol. 1, 1972; vols. 2 and 3, 1973; vol. 4, 1974.

Order from: Q Corporation
           49 Sheridan Ave., Albany, NY 12210

Index

ABOUT THE AUTHOR

Frank Rosa, a field engineer with the Metropolitan Refining Company in central New York, has been a water treatment professional since 1961, doing lab work, drawing up proposals and quotes on new construction, overseeing quality control, and serving as a treatment consultant to engineers, architects, and mechanical contractors. His articles on topics such as water treatment hazards, poor design in steam boilers, algaecides and biocides, and boiler flooding problems appear frequently in professional journals.